How Katie Pulled Boris

Travels with an American Motorhome (RV) in Europe

Keith Mashiter

Published in 2013 by FeedARead.com Publishing –
Arts Council funded

A CIP catalogue record for this title is available from
the British Library.

First Published 2007 by
Exposure Publishing,
an imprint of
Diggory Press Ltd

Cartoons and Cover design by www.tonystoons.co.uk

Rear cover photo by Roger Coy

Thanks and deep love to my wife Gail who pretends to tolerate my maniacal ideas.

CONTENTS

You Want to do What?

'Where did this idea come from?' asked my dear wife Gail.

'I've no idea,' I responded.

'But Keith we've absolutely no experience, the last time we did anything like that was when we had the Volkswagen pop-top in America and that was over thirty years ago and if you remember we broke down on top of the Rockies and had to be towed for twenty miles.'

'Well I wasn't thinking about a pop-top; we would want something more comfortable like one of those American Winnebago things; they've got everything on board: you get a large bed, lounge, toilet, shower, kitchen, those fancy American fridges, and TVs.'

'Keith you've always said there are great last-minute travel deals on the Internet and when we retire we'd be able to take advantage of them.'

'Well I did, but who knows what you get with those cheap deals; do they change the sheets let alone the bed covers, you may find pubes in the shower, and then you're stuck in one place and have to hire cars or go on coach trips with all those people you've tried to get away from? No, with the Winnebago you can choose mountains one day the sea the next and if you don't like your neighbours move on. You carry everything with you and you know it's clean. I've got some pictures of one from the Internet; it's not a Winnebago but very similar.'

'But that's huge and made for wide American roads, God knows what the fuel consumption is and where are you going to park it? Remember, we live in a London flat.'

Gail was right, it did look bloody enormous. Would I be able to drive something that big, did you need a special licence? There must be somewhere to park it – maybe I could find a truck park or something similar – what did other people do?

'And don't you need some transport when you get there; how would we get around to do the shopping or go sightseeing?'

'Well we could tow a car behind. I've seen them do it in America.'

'But our S-Class is huge as well, I don't see how you could tow that, the length would be enormous. Anyway is that legal over here and do you really want to be seen driving something with '*Georgie Boy*' written all over it?'

It was true we had no experience of camping, caravanning or motorhoming apart from the episode a distant thirty years earlier with the pop-top. That had also been a spur-of-the-moment decision.

We were a newly married and hard up couple spending three years working in Pittsburgh USA. We managed to scrape enough money together to buy a second-hand green

Volkswagen Beetle – a wise choice for an approval-seeking new son-in-law knowing that Gail's military father had revered the one the family owned whilst stationed in Germany. Every spare weekend and holiday we seized the opportunity to travel somewhere different. We toured Pennsylvania and West Virginia State Parks and camped in a borrowed tent.

When I visited the local Volkswagen dealer to get the car serviced I passed the time in the usual manner by wandering round the showroom critically appraising the new and totally unaffordable cars. Taking pride of place, rotating on a raised turntable in the centre of the display, was a striking red, pre-owned Volkswagen pop-top caravanette. Its sliding side door had been left open to show the interior wood cabinetry and ochre plastic seats all made to look remarkably spacious because the elevated white plastic roof provided a canvas canopy and headroom to the centre area by the door. I peered inside; it had a cooker and cold box, dining table and the bench seat and rear area made up into a double bed.

With time to kill and my parents visiting us in the USA for the first time, I started to form the idea of travelling so we could show them the forests, mountains, and rivers we had come to enjoy so much. They could sleep in the pop-top, we in the tent. The price of the pop-top was more than we had ever paid for a car and we didn't have the money so Gail probably thought it would never happen. Caught up in my own enthusiasm I approached Mellon Bank on Forbes Avenue with some trepidation, as I had never borrowed before, not even a mortgage. The heavy chandeliers, the dark wood desks, tiled floors, marble pillars and the polished brass grilles of the cashier's booths made the banking hall seem particularly cold, cathedral-like and austere. Without an appointment, I was ushered into a heavily panelled office to discuss the loan with the duty manager. I put my quickly thought up case to the button-down-collar, white-shirted, crew-cut, all-American sports looking bank official: no savings, temporary job, short-assed non-resident visitor, with long hair and weedy Zapata moustache, thought it would be good to spend a load of money we didn't really have.

'Hey! No problem – how much would you like to borrow?' and the form was signed.

'Have a nice day and enjoy your new automobile.'

There had been less formality than trying to buy a bottle of wine at the Pennsylvania State liquor store where on my first visit they had dismissed my UK driving licence as proof of age, thereby embarrassing me in front of the queue of alcohol-deprived Pittsburgh steel workers. They then telephoned my new employer (the University Professor of Medicine) to determine whether I was old enough to be out on my own (I was 26 with a PhD). On request for a recommended wine they stuffed me with an Upper New York State white wine that tasted like I imagine horse's urine would and probably had a good laugh about it with the bourbon purchasing roughnecks.

We received the resplendent red pop-top from its inspection and service only on the morning my parents were scheduled to arrive at Pittsburgh airport and with no time to spare, but great anticipation, hurried to the airport. That was the first time I had driven the pop-top and I revelled in the unaccustomed higher driving position, the unique sound of the rear air-cooled engine and all that interior space. What a surprise it was going to be to my parents who had no inkling what awaited them but probably imagined American limousine and penthouse style living now that their only son had a job in that country where the streets were paved with gold.

On arrival in the airport car park I proudly descended from our gleaming red transport and looked around secretly hoping to see some admiring faces amongst my fellow Cadillac and Chrysler drivers. It was then I noticed the trail of oil along our entry route. Surely that had to be from an earlier vehicle? No, it was ours. With parents arriving imminently all sorts of schoolboy chiding following the purchase of my first car flashed through my mind: 'What did you think you were doing buying that thing?' Gail was strangely silent, what was she thinking? I couldn't find any soothing words. I was worried enough myself. We met my parents and ignominiously rode back to our apartment in a yellow cab.

The dealer recovered the vehicle; somebody had failed to tighten the sump nut in the hurry to complete the service. We got the pop-top back and we all travelled in it to upper New York State, The Adirondacks, and Niagara Falls. The encounter with the black bear behind the picnic table and a disbelieving mother still brings a smile to my face but looking back I can't remember how we, the tent and all our gear fitted in or how we managed with a two-burner stove and an ice-box in a vehicle that was no longer than a Ford Mondeo. What did I put my parents through? But that same pop-top then took Gail and me out every weekend, and on memorable vacations down the Blue Ridge Parkway (Cherokee Joe are you still making the best pizzas we ever tasted?) and 2,000 miles across the USA central plains to Denver and up the Rockies. There, in the last 30 yards of a near-vertical climb the engine expired in a cloud of blue smoke.

Immovable in the middle of nowhere we hitched a ride with horsebox-towing Texans in a truck that seemed to have room for eight people across the front bench seat, along with the crate of beer. In Durango we rented a U-Haul truck and tow bar and went back up the mountain to retrieve our abandoned pop-top. Having towed it into Durango we were directed to a garage that had only opened up two weeks previously. According to the owner of this new and strategically positioned Volkswagen repair shop, almost all Volkswagens, unadjusted for the atmospherically weaker oxygen content at high altitude, gave up the ghost. He congratulated himself on the flying start to his new business plan as he anticipated the three days work required to strip the engine down and replace the valve that gasping for air in the rarefied atmosphere, had been sucked through and embedded in a piston.

Our vacation was on hold so we settled for the local motel and mingled with the cowboys at their local show. We then enjoyed the rest of the trip to Mesa Verde, the magical Bryce, and Zion National Parks and thankfully drove home without further incident using our red pop-top for many subsequent outings. But having left the USA we never went near a tent, caravan or motorhome again.

I don't know where the idea came from to buy a big American RV, tow a car behind and bugger off for months on end around Europe without a care in the world. Park up, enjoy the luxury, the views, go walking one day, join the smart set in St-Tropez another, be who you wanted to be when you wanted. It sounded so simple and idyllic.

I remember sitting in Costa Coffee at the services just north of Cambridge nursing a cappuccino having got thoroughly rain-soaked on the way in. I sat at the window counter musing on life in general, the ordered grind of it all and getting some perverse satisfaction watching others dancing over the puddles in their haste to escape the downpour. I was filling time prior to an appointment with a client, one of a breed of Cambridge technology entrepreneurs who had suddenly made millions on the back of an innovative idea and a start-up company. I had moved from medical research into the exalted world of city finance. My client, and that was stretching the term in the interests of my monthly statistics, still preferred a bicycle to a car, rented accommodation to an owned property and had put most of his money in National Savings except for the £7,000 he had entrusted to our fund managers that I noted from my report was worth £5,800. The share markets had gone into terminal decline. I looked at our fund manager's glossy report *'Buy now it'll never be as good as this again, shares have always made money over any consecutive five-year period'* with many coloured graphs to emphasize the recoveries from world wars, Pearl Harbour, tulip blight, and cataclysmic disasters from time immemorial. *'Use this positively with your clients to secure further investment, get them to see this unheralded opportunity.'* Had they ever met a losing investor face to face, like I was about to? How depressing could things get?

I wandered into the shop to buy a *Financial Times (FT)* looking for something deeply comforting that I could quote. Scanning the shelves I saw a sparkling white motorhome parked by a sun-drenched Mediterranean on the front cover of *Motorcaravan and Motorhome Monthly' (MMM.)* I languidly flicked through the pages then put it back on the shelf, better get the *FT*. But had a seed been sown?

Shortly after the motorway ruminations, an unexpected offer to buy my business practice provided an opportunity to retire early, escape the proverbial rat-race, albeit on limited means and brought the motorhome travelling idea back into consciousness.

Gail's scepticism may have been based on an inherent fear of mechanical things breaking down. It wasn't only the pop-top. Long before that I had taken her out to a black-tie dinner, she resplendent in a full-length, pink and grey jersey dress, I in a dinner suit with an engagement ring in my pocket. After dinner we drove out of the car park in my outwardly impressive but inwardly unreliable silver TVR sports car eliciting a crunching sound underneath. I climbed out to see what might have happened.

'It's no good looking under there; it's over 'ere!' a broad North Country voice rang out in the dark.

The full length of the exhaust pipe lay in the car park. I scooped up the rusty object and shoved it through the window to rest one end on Gail's lap. We drove home with the other two-thirds sticking out the window like a gun turret, but we were engaged.

Well Mickey Likes It

'Howdy, folks. I'm Bret. How y'all doing today? Wanna check out what we've got? Jump aboard.'

We were in Orlando, Florida. Mickey Mouse land has more motorhome (or RVs as I was learning to call them) sales outlets and keener prices than any other US State. Where better to look at American RVs? Of course getting there had required some subtlety on my part.

'There's this great last-minute Internet deal to Orlando; it will be a good opportunity to follow up on your idea for our new retired lifestyle and get some winter sun, what do you think?'

'Well Orlando wouldn't be my first choice. You're not thinking of going to theme parks or something daft like that, are you?' Gail had queried.

'No, of course not,' I lied.

After arrival and sneakily consulting *Yellow Pages* in the hotel room I persuaded Gail to ignore the puddle the hotel called a pool and go with me on a tentative, just for a look, as we happen to be here, drive to the western side of Orlando where all the RV dealers seemed to be adjacent to each other.

Bret dazzled us with his super shiny grey suit, black boots, cream coloured cowboy hat, and silver necktie. We jumped aboard his golf cart and were whizzed around the lot that was full of RVs.

'Hey! Watch out for the steps – they'll come out automatically when I open the door,' he said, showing us into a 36-footer.

'This here's a medium sized RV we go right up to forty-three-footers. Here you've got a three-seat couch that converts to a bed, a booth-type dinette for four that can also make a bed for two, a deep pile carpet then a fully fitted and tiled kitchen with dual sinks, hob, oven, microwave, and full fridge-freezer. Moving through there is a bathroom with shower and then here in the bedroom we have a full walk around queen-sized bed and closets. So you can sleep six. How many kids you folks got?'

'No, kids.'

'Well that's a nice amount of room for you two and you'll have air conditioning, gas heating, hot-water heater, TVs, video, CD multiplayer and an electricity generator; everything you folks are going to need and right up front you've got that great eight-litre Chevy engine and Allison transmission.'

I sat in the mock leather driver's seat, gripped the steering wheel, handled the automatic gearbox lever, and peered through the enormous windscreen. I looked out of the side window at the six-foot drop to the ground then I swivelled round to look into the spacious lounge. Bret pressed a button on the wall and the side of the motorhome containing the dinette unit and sofa, in other words an open box about 12 feet in length and 3 feet in depth slid outwards and an extra 36 extra square feet of floor space materialised tardis-like.

We looked at a couple more RVs, equally sumptuous.

'You folks aiming to buy today?'

'Well, not right now.'

Bret had realised we were not going to make his sales figures go stratospheric and spotted another couple who might, leaving us his card.

'Howdy, folks. How y'all doing today? I'm Bret. Just had some folks that came all the way from London England to see us. Now where are y'all from?'

'Orlando, well that's great, good of you to stop by. Jump aboard,' and off he went.

At the lot across the road Henry met us as we parked the car. He was also in a grey suit but with a conventional blue tie and looked like everybody's favourite uncle.

'Hi, folks, I'm Henry. How y'all doin' today?'

'We want to buy a motorhome to export back home.'

'Have just the thing right here for you folks – a new thirty-foot Itasca Sunova with an MSRP (Manufacturer's Suggested Retail Price or sticker price) of eighty-thousand but I'm sure you don't want to pay that so we could let her go right now for sixty-four thousand. We can ship her out through Jacksonville and you won't have to pay the six-and-one-half percent Florida sales tax. What d'ya think?'

Wow! We were being offered a brand-new motorhome, with a pleasing light interior, from a well-known company, (Itasca are the upmarket part of Winnebago), for £40,000 and at a 20 per cent discount. I was sure these were selling for close to double that at the UK dealer. What would shipping be, would taxes have to be paid? Gail said she quite liked it and we might have gone ahead there and then but something was holding me back (like still knowing sod all about the things, the import costs and having nowhere to put it when it landed in the UK).

The next day we set off west in the rental car for Clearwater. I took great delight (but perhaps to someone else's annoyance) in pointing out every RV we were passing on the road, all going surprisingly quickly despite towing a large Jeep Cherokee or similar. I began to imagine what it would be like cruising down the highway in our own RV. Gail was probably thinking the same but imagining Bank Holiday on the A30.

Half way to Clearwater we spotted hundreds of RVs off to the right and quickly left the road to try to find them. After making some enquiries from locals we eventually came to a 300-site campground chock-a-block with RVs and then a magnificent modern sales centre with a pillared entrance like a Dallas residence. By chance we had arrived at Lazy Days, the world's number one RV dealer with annual sales income of $700,000,000 and 1,250,000 visitors served by 600 employees including 150 sales consultants riding 155 six passenger golf carts to cover the 1,200 RVs on sale. It was mind blowing. Round the back 250 technicians in 220 service bays could repair or service your RV, change your interior furnishings, or airbrush murals to make your RV unique. This was RV city.

We hurriedly walked through the sales entrapment area that looked like a computerised city-dealing room manned by hotshot-suited salesmen. We gave the once over to some of the RVs on display in the huge park-like grounds but by then were getting hungry so bought beef sandwiches from a stall thereby missing out on the restaurant that serves over 430,000 complimentary breakfasts and lunches to customers and prospects annually. It was too much; after only ten RV visits my eyes were beginning to glaze.

We visited other dealers but none compared with Lazy Days. It looked like we might get a real bargain if we imported from the US particularly on a new vehicle even after shipping, insurance, import duties, and VAT. Why were the UK dealers' prices so high? But was it all going to be worth it and what if something went wrong, we could hardly take it back to the dealer? Seeking enlightenment I went browsing in the aircraft hangar otherwise known as *World of Books*, and found just what I was looking for Brent Peterson's, *The Complete Idiot's Guide to RVing.*

A Wide Boy

'I'll go along with you whatever you decide,' Gail said, which any man will know as – it's entirely your fault if this goes wrong and I will remind you of it for ever.

I was choc-a-block with vital information from our American trip, UK magazines, and the Internet. I knew that a doghouse was not for your pets but the engine hump that stuck up into the cab. A CAT pusher was not pet-related but a make of diesel engine that was usually installed in the rear to push you along, hence a diesel pusher. I wasn't sure whether the car I towed (*'toad'*, gettit?) would need a brake buddy to help it stop, and should be on a tow dolly or all four wheels down. It didn't take an intuitive leap to realise that black (toilet) water was going to be dirtier than grey (shower) water although I would have thought brown more appropriate. A split bathroom wasn't one needing repair but one that had a separate toilet. A barrel chair not some brewery recycling activity but its shape. Hook-up had nothing to do with hooks but plugging in the RV to shore (campsite) electrical power and a dump station was what you needed for the black and grey stuff when you hadn't room for any more.

We visited UK dealers up and down the country; most had fewer than thirty vehicles and paled in comparison with Lazy Days where the 1,200 RVs were presented immaculately.

'We hire these to Formula One drivers,' enthused one dealer.

Well I was excited at the prospect of enjoying the same comforts but Gail didn't seem as interested in sleeping in the same bed as David Coulthard or Michael Schumacher she was simply hoping this was another of my passing manic phases. We would often step into a vehicle with a garish coloured sofa and chairs. Many had the country cottage look with strong colours and flowery patterned velvet fabrics, bright pink or maroon chairs, mirrors, and lockers with scrolled patterns – we preferred a contemporary look with beiges and cream leathers, however it may be for others and far more practical where children or animals are concerned. Some were dirty and looked like they had been out on hire to builders; others although nearly new stank and obviously needed a visit to the dump station.

At one dealer we deigned to look into a glittering display of upmarket £250,000 vehicles that had a jobsworth at the door.

'Hello, we'd like to have a look around.'

'Are you in the market for a coach?' (Note the word coach – an upmarket RV).

'Yes we are,' I said more tentatively than it appears.

'You realise how much this one costs?'

I wondered why he was asking this, as the price was printed in big black letters in the window.

'Yes I do,' I said meekly and after making sure we had wiped our shoes we were allowed entry to a palace of cream leather, Corian worktops, planked floors, illuminated ceilings, mirrored walls, gold-plated showers, surround sound, and matching drapes. After the doorway interchange it felt like we were trespassing.

We met with Patrick who was selling a 37-foot RV that did have a cream leather interior as well as green moss stains coming off the roof, bird droppings on the wing mirrors and a large *LEFT HAND DRIVE SIGN* on the rear quarter covering a dent in the fibreglass.

'This is the scary bit,' said Patrick as we took our first ride in an RV and he did a three-point turn across a busy main road with vehicles hurtling round a blind corner. On the open road we were soon up to 70mph settling back smoothly on cruise to 65mph.

'Once a glass fruit bowl shot forward off the kitchen counter jut missing my head,' he confessed.

I looked up at the TV above his head and held in by gaffer tape.

'It's, very nice, Patrick. We'll give it some thought,' I said over a coffee afterwards.

Matthew was young and drove a black Porsche and we arranged to meet outside a pub (your thoughts are the same as mine were). He was selling the brand new 32-foot RV squashed against the hedge in the front garden of his seaside bungalow. It had swishy exterior graphics and was surprisingly cheap. As we stepped inside, Gail blanched when she saw the interior colours of Scottish Plum and the front seats in a corduroy version of Harris Tweed.

'The interior's a horrible colour – just look at those chairs, I don't like it,' she said.

'Look at those window drapes – instead of being beautifully bound they're hanging down loose and thin. There's no room between the sink and cooker to put anything.'

From my point of view the internal arrangement was fine; it wasn't quality as we had seen before but it had everything in the same order as many a more expensive RV. It was new and we could afford it.

As we drove away, Gail and I discussed the remarkable value or rather I put forward the case and she argued

'It was hideous. Keith, there is no way I could live with that and I don't know why you're even thinking about it.'

Perhaps she wasn't quite won over yet but at least we were now discussing the merits of individual vehicles that we might buy.

I checked with the main dealer for that make of motorhome.

'And a new one would cost?'

'Eighty-eight thousand pounds.'

'I'm being offered one at fifty-eight thousand.'

'Ours has a warranty unlike the back street dealer and has been built to European Standards with a front-engined diesel and a higher specification.'

'What is better in the specification?' I asked.

'It will have double-glazed windows, a generator, air conditioning, fridge freezer, and leather front seats,' he said.

Matthew's had all of these except the diesel engine, leather seats, and double-glazed windows. The leather seats cost an estimated £1,000 so £29,000 for a diesel engine, double-glazing, and a warranty sounded a lot.

What a bargain I had found, but Gail still didn't like the interior and inwardly I knew this was not the high quality vehicle we craved, I was up another blind alley. Should we import from the USA? They had all looked so much nicer in that Florida sunshine.

A number of companies and people offered to convert imported RVs. They replaced the TVs and videos and put in standard British 240v electrical sockets; some said the fridges and microwaves had to be converted and they changed all the vehicle lights. Customers with unlimited finance could have solar panels fitted to charge the batteries, satellite TV systems or GPS navigation for around £1,000 each, and if I chose a petrol engine it could be converted to LPG for £2-4,000. All sounded wonderfully professional.

Afloat in this sea of advice and unknowingness I was desperately looking for a lifeline. I phoned East Coast Leisure in Basildon and spoke to owner Nigel.

'Many of my staff and I own RVs. We regularly buy for customers from Lazy Days at good prices and can ship for around three-thousand-five-hundred pounds and convert for two-thousand-five-hundred.'

I liked the sound of Nigel, a real Essex bloke; perhaps he would be my lifeline?

'Have you thought about the *width* of the motorhome?' he asked.

'Well yes. I know it will be big.'

'No, the maximum legal width in the UK is one-hundred point four inches and few American RVs comply with this.'

Vehicles were on the roads that were illegally wide? Surely he couldn't be right, how could they be sold? I found it hard to believe; perhaps he was spinning me a line, was he the wide-boy? I called the dealers we had visited previously to check with them. Whereas they were silent about it before, they were now quite forthcoming about the width regulation as well as their compliance with it. But how was it they apparently had vehicles that were too wide on sale? Their explanations were many and to my mind somewhat improbable:

'Another company imported it and insisted these models were legal.'

'The manufacturer includes awnings and handles and various other components in the measurements which are excluded by the UK authorities.'

'The UK regulations exclude measurement of the flashers on the side but the manufacturer does.'

'UK models have wings fitted without flares.'

'UK models have a different chassis.'

'UK models are modified in the US factory and a new front end put on but I'm not sure how they do it.'

Whilst apparently being whiter than white in their own dealings they unvaryingly referred me to transgressors at other dealers

'I just don't know how dealer X is doing what they are doing – their models are clearly in breach of the regulations.'

'We refused to take a trade bought at X because we knew it didn't comply.'

'Some dealers' models are not complying, whereas all of ours do.'

I emailed US manufacturers directly.

From: X Corporation: All of the units X builds for 2003, or the 2002 models are 101" wide. This does not include the awnings, mirrors, or grab handles. Thank you.

And later:

> *Have you spoken to our dealer in the UK? He is buying coaches and imports right now. Maybe he can find what you are looking for, or give you some hints on importing the unit?*

The same dealer who insisted these models were 99 inches and awnings were included.

> *Hello Keith, Thanks for your interest in Y motorhomes. We measure the width from sidewall to sidewall. This means that we are about 102" wide considering moldings, etc. You will need to go back to a 1997, 1998, and 1999 model to get a 96" model. We did offer optional wide bodies, or 101" wide models during these years. Please be careful when purchasing, and be sure that it is a conventional width coach.*
>
> *We have a dealer in England who sells all our product lines. Thanks again.*

I replied that I was aware of the dealer but concerned how they import under the regulations – they say they have different wings on the vehicles but can this really be so?

A further message from company Y:

> *They have been doing it for years. It appears that our coaches are within tolerance in the opinion of the dealer.*

I searched the US manufacturers' web sites for the width dimensions and found almost all did not comply because the competitive trend in the USA was to go with wide-bodied motorhomes and many were 102 inches. I was mystified; had I stumbled upon some silent conspiracy? On the Department of Transport's website I found *Booklet P15 – How to Import your Vehicle Permanently into Great Britain* and it confirmed that vehicle maximum size was a length of 12 metres (39.37 feet) and width 2.55 metres (100.4 inches).

A representative at The Vehicle Inspectorate said that more experienced colleagues at Vehicles Standards and Engineering were aware that the rules were being flouted. People could still import and register over-wide vehicles with the DVLA. The problem would be if they were pulled over by the authorities such as the police when they *could* be prosecuted. I looked up *The Road Vehicles Construction and Use Regulations 1986 (No. 1078)* and under *Overall width Regulation 7/963* it said that mirrors, lamps, reflectors, sideboards let down for loading and usefully snowploughs were not included.

My confidence in UK dealers was at low ebb. It appeared legal to sell a motor home that was too wide but illegal to own and drive one and a court case had come to the same conclusion, driving that RV owner into bankruptcy and hanging out a big red '*Caveat Emptor*' flag for all RV purchasers. I wondered what would an insurance company do if you had an accident and were driving an illegal vehicle? I began to feel we were not going to find what we wanted at a price we could afford and could get duped and end up breaking the law. Did we really want that worry every time we saw a police car as we were driving along? It was hardly the scenario I had envisaged; was the dream at an end?

I called Nigel to unburden my depressing thoughts.

'I think I may have what you need. I've just had instructions from a client to sell a Holiday Rambler Vacationer. It's just over a year old, low mileage, never been slept in, and the covers are still on the carpets, so it's like new. We imported it and did the conversion. It's thirty-three-foot two-inches long with a kitchen-lounge slide out, the 6.8

litres V10 310 HP Ford Triton petrol engine, generator, levelling jacks, twin TVs, video, air conditioning, Cat 1 alarm, and loads of extras. You won't get a better one.'

From a mood of giving up it was hard to turn back onto an enthusiastic track. Our ideal was a Vacationer but new ones were going to cost over £90,000 and this was at a substantial discount. We went over to view the Vacationer. The bottom was painted metallic green, the rest white with curvy stripes of metallic gold going to green.

'This is nice,' said Gail, as we walked in.

'Look at the cream leather chairs – that's what we wanted. I like the pale green patterned sofa and dinette and all the drapes match and look at those plush window dressings.'

'Can't go wrong with a twenty-inch TV,' I said as I tried the swivel recliner chair for size imagining myself parked up with a mountain view out the windows and footy on the tele.

'Have you seen the size of this fridge and there's a big freezer too,' Gail continued 'and I've never had a microwave that big.'

'Now this is quality,' I said, feeling the oak cabinetry of the kitchen and lounge.

At the rear was a large wardrobe; a walk-through bathroom, with a vanity unit, a large shower, and a separate enclosed flush toilet. The bedroom had a walk-around queen-sized bed with covers that matched the curtains and window dressings, shirt closets, and nightstands, a TV and mirrored headboard.

It had everything we could want as options – day-night blinds, multi-CD player, VCR, rear vision camera, driver's door, 50-amp generator, two air conditioners, dash air conditioning, in-floor ducted gas heating, 8-gallon water heater and hydraulic levelling jacks. The kitchen top was not

Corian but nicely finished and the space between sink and cooker was not large. But it felt right. The 6.8 litres V10 engine was started and it sounded like a sports car; a deep throbbing hum coming from the three-inch exhaust.

Some serious negotiation on the price and we had bought our RV. Not that we had driven it or been for a ride anywhere we simply assumed it ran reliably. Nigel had promised he'd make sure everything was in apple-pie order and even provide parking for us.

I eagerly sent off for the 2001 brochure:

Subject: Holiday Rambler Vacationer 2001- I sent your information out! It should be coming soon. Sincerely, HR Sales.

It looked great and it confirmed we hadn't bought a (very) wide boy at 100.5 inches.

I phoned an insurance company to get cover.

'That vehicle is over 7.5-tonnes. Do you have an LGV licence?'

'I regret we won't be unable to insure it until you do.'

Big Yellow Truck

'The computer doesn't always recognise your mouse click so it's probably best to give it a couple or so to make sure,' advised the official at the grey and blue gulag otherwise known as a Driving Standards Agency test centre as I stooped gratuitously at his little window. Another government computer project successfully completed, I thought as I entered the darkened room with banks of computers.

'But don't do more than six or you will be eliminated,' he shouted as an afterthought.

From the two DSA tomes about driving goods vehicles and official theory test bought in preparation I had learnt the tachograph regulations, how to tie my load down with 'dolly knots' and when to ask for a police escort for my over-wide RV. What I didn't have was any information on the new Video Hazard Perception Test, as with the usual bureaucratic efficiency it had not arrived in time. As the count-down to the first video started I began to panic, as I couldn't even recognise what the static picture was of. Then a ghostly shadowy figure passed across the screen. Was that it, had I missed the hazard? The indecipherable picture then manifested as the back end of a tram moving away. How appropriate is that to modern day driving I thought? But what was the ghostly figure?

The fourteen videos passed quickly and I saw potential hazards everywhere zapping them with my mouse like a shoot-'em-up playstation game, but not more than six times, of course. Who had thought this up? Computer feedback requested at the end ranged from 'I thoroughly enjoyed participating in this well thought out and challenging programme' to 'I was disappointed this wonderful test didn't go on longer' but did not include my preferred 'Do you realise this load of crap counts for 40% of my test result?' I readied myself to give the official a verbal onslaught.

'Very good, full marks on the theory, a little less on the hazard perception but a pass overall.'

I thanked him and left. Who cares about the videos? I'll never have to look at them again. As I drove home I pondered the need for the LGV licence. Some American motorhomes, based on the same Ford or Workhorse chassis, the same size as ours with a slide-out and similar internal layouts and appliances, were advertised 'drive on a car licence.' Was the quality of their house construction so flimsy that their motorhomes were lighter? This was, and remains a mystery to me. Some sellers of heavier RVs had suggested I would never be stopped, it was a grey area, the police would not know, turn a blind eye, no court case had ever been brought or, that the vehicle could be down rated to comply.

Our Vacationer unladen weighed 7.7-metric-tonnes, tempting a down rating to the magic 7.5-tonnes when an LGV licence wouldn't be required. However, that would mean we couldn't carry Kate Moss let alone her wardrobe, whereas it was made to carry 1.9-tonnes of stuff including water, propane, food, and all the people who will eat it plus their gear giving it a Gross Vehicle Weight of 9.3-tonnes. I had to get an LGV licence or, be a 7.5-tonne (fully laden) lawbreaker, and as the insurance company said they wouldn't start cover until I had one there seemed no other option, and to be fair to Nigel he had always recommended I get a licence.

On Tuesday May 6, I started driver training at an aerodrome; the fee was £663.50. I liked the idea of not driving on public roads until ready. In the long grass at the side of the airport perimeter road was a scrapyard of derelict vehicles, moss-covered caravans, and a rotting portacabin with blue tarpaulins protecting the roof, where I met Jim, my instructor.

'Where's the truck, Jim?' I enquired.

'If you can drive these you can drive anything.' Jim laughed as he climbed aboard a twenty-year-old Bedford box van for my lesson.

We set off out of the long grass up the aerodrome. The gearbox was vague, the steering heavy, and the engine wheezy and polluting. In contrast the brakes were unexpectedly sharp and I nearly shot Jim through the windscreen. Jim showed how the reversing exercise would be done at the test centre and I achieved it first go and felt pretty pleased. We set off for 4 hours driving on the road. My shoulders were soon aching, power steering apparently not being an option twenty years previously, so I was relieved when we stopped halfway at a greasy spoon for an unaccustomed egg, beans, and chips. Jim lost interest in his after finding a stone in the meal.

Back at the airfield we watched two other trainees reversing with an articulated truck.

'Jim why are they doing that differently to how we did it?' I queried, knowing that what we had done was much easier.

Jim mentioned this to the other instructor and a heated discussion started from which it was clear something was wrong.

'It's because their vehicle is articulated,' Jim finally explained unconvincingly (all vehicles are tested the same way).

'Look, Jim, this twenty-year-old truck is really hard work – is there a better one?' I asked.

'There is a newer one and the gearbox is definitely better but the steering poorer. You can have a go in that if you want,' he said evidently relieved not having to explain about the reversing anymore.

'How old is that one then?'

'It's nineteen years old.'

I retrieved it from the grass and drove it up the airfield road and back down again.

Jim was standing with the lady in charge. 'You can't use that one after all, it's not roadworthy.'

Overnight I didn't sleep and worried about how their vehicle and organisation were reducing my chances of passing. The next day I drove over to the aerodrome.

'I'm quitting,' I told Jim 'the vehicles are a joke.'

'You're dead right. I've been telling the owner to get new ones for years. You should try the training company in Chelmsford.'

The secretary agreed a refund of the money for the days not taken.

I had a test drive with another company in a big yellow modern DAF box van with power steering.

'It was a smooth drive,' Chris the young instructor was kind enough to say as he took the £671 cheque for me to be able to start lessons the following Monday. When Monday came he told me to forget my driving technique and start afresh. We spent four days driving round the high streets of Enfield, Chigwell, Waltham Abbey and Forest with nary a dual carriageway to be seen. By day two I was not sleeping, thinking ahead to the driving, the test, and the pressure to pass so we could get insurance cover. As we peered over the fence at the test centre I watched someone botch the reversing test confirming the many horror stories of failures and reported 50% pass rates for first-timers. By day three my driving was worse and on the fourth day the pressure increased as my eighty-year-old mother arrived to stay, unaware of our RV purchase and plans to leave her and disappear off abroad for long periods.

'Chris what are we doing?' I asked.
    'I don't like to arrive too early at the test centre as other instructors wind candidates up.'
    My prophylactic Imodium taken in readiness for the Friday ordeal was being tested to the limit. In the waiting room I sat, stood, looked out of the window and sat again waiting to be called. Gary, smartly grey-suited and all togged up in fluorescent jacket with a bundle of papers on his clipboard, politely introduced himself as my examiner. The paperwork done we went outside and he asked me to bring the vehicle onto the manoeuvring area and in so doing I

thought I had run over his foot. I completed the reversing test.

'Keith, are you satisfied with your position?'

No time for jokes, I thought and so I said I was. The back of my truck was within the one yard allowed. We went on to do an acceptable braking test then set off from the test centre into heavy traffic, crawled, then eventually were free and onto some narrow roads.

'Keith I see you have Doctor in front of your name. Why has there been no cure for cancer yet?'

'It's complicated, Gary.'

'Now take my diet,' which he explained in detail, 'would you regard it as healthy?'

'Hard to say, Gary.'

'I'm most worried about not being able to stop smoking,' he continued 'what would you do – are there any quick cures?'

Get a less stressful job, I thought, struggling to drive down Chigwell High Street with footballers wives' Porsches double-parked and with Gary's constant questions and dietary requirements my concentration was lapsing.

'Gary, I thought I was coming for a driving test, not a medical consultation.'

I hoped I hadn't said it too sharply. The test seemed to go on and on. I peeked at the dashboard clock and tried to work out how long we had to go. I then realised we were on the road back to the test centre.

'So why do you want to drive a truck, Keith?'

I explained the RV and this generated more questions – was this a good sign? Would he have chatted with me in this way if he were going to fail me? Finally we were in the test centre and I switched the engine off.

'There's something I need to discuss with you, Keith. On one of your manoeuvres you didn't look into your blind spot but it's not a failing offence so I'm pleased to tell you it's a pass.'

I had passed a test in a right-hand-drive diesel box van with manual gearbox and was now entitled to drive my left-hand-drive petrol RV with automatic gears and rear-view camera. I called Gail and drove home – it was Yorkie bars all around. I was a truck driver. We could insure the RV and were ready to go.

Boris of Toad Haul

Touring with a 33-foot long, 8-foot four-inch wide, 12-foot high RV was not going to be convenient for popping into the local supermarket for the weekly shop, exploring unknown country lanes that might turn out to go nowhere but *Jean-Philippe's* farm or enjoy an evening out at a bijou *Provençale* restaurant. We had seen RVs in the USA towing large vehicles behind without a trailer and 'all four wheels down' delightfully called a 'toad' or 'dinghy' for those of a nautical persuasion.

With this in mind, our large Mercedes S-Class car was sold in favour of a white Mercedes A-Class A140. I kept looking at it in the underground garage to our apartments; it was half the size of the space previously occupied by that great steel status symbol. My life had changed. Gail thought it was cute and even named it BORIS (Benz). I think she was glad the big S-class had gone as it brought back memories of our last outing up the M6 when a white transit van had ploughed into the back of us as we sat stationary and helpless in an outside lane traffic queue. Our lives were saved by the strength of the Mercedes and we drove away unharmed with a dented boot whereas the front of the van was demolished.

Boris was short, light and yet had a huge load carrying flexibility as all the seats, bar the drivers, could be moved, folded or removed. We needed this to transport all our stuff out to the motorhome. Boris grew on me, the seating position and motorway speed didn't make me feel like I had descended too far down the motoring status chart and I was confident on the safety front should any elks dash across in front of us.

Although I hadn't given it much thought before with the shenanigans over RV widths, I started to worry about towing Boris behind the motorhome and UK law. Once again this appeared to be a grey area and advice was equivocal in two respects; the towing of a car on its own wheels without a trailer with what was known as an 'A-frame' and the driver's licence needed. Many towed on A-frames without any further thought on either topic. An A-frame locked on to the motorhome's tow ball like a caravan and then from this two bars were attached to a horizontal bar on the car. Once attached safety chains and the wiring harness were connected and the cars ignition turned to a point where the front wheels were free.

I learnt that a vehicle fitted to an A-frame became a trailer in its own right and should be legal provided it complied with the lighting and braking requirements. The motorhome's signals and lights operated the corresponding ones on the car and with red reflector triangles fitted lighting was covered. When the motorhome slowed the car behind pushed against an inertia sensor on the A-frame and pulled a cable attached to the brake pedal applying the car's own brakes. The problem came if the RV was reversed since the same mechanism would apply the brakes whereas EU law required that such a mechanism be reversible without having to get out to detach it. Yet millions of Americans towed with A-frames.

It seemed less of a problem in the UK than in Europe and yet Nigel had told me he had towed all over Europe with an A-frame and correspondence in an RV magazine showed many others were doing the same. A manufacturer of A-frames cited a case where someone had been stopped by Spanish police and fined, but on appeal Brussels had revoked the charge. We decided that the convenience of an A-frame was worth the risk compared to using a heavy trailer that needed to be parked on a site along with the car and motorhome.

I was depressed to find the law on licences clear. If the motorhome exceeded 7.5 tonnes and the kerb weight of the car exceeded 750 kilograms (and most did, including Boris) there were two consequences. I needed to hold an 'E' licence in addition to the 'C' licence previously obtained. The only way to get this licence was to take a further driving test in an 18-tonnes articulated truck or a vehicle in excess of 10-tonnes towing a suitable trailer. I had to go back to the driving school; were we ever going to get going? Roger said he could fit me in on June 9, for a five-day course but this time I insisted on afternoons and against Roger's advice wanted solo lessons.

Abbey Wood, Would Katie?

Having started the naming game with BORIS, we decided to call the motorhome KATIE from Vacationer. It made talking about them much easier as we still didn't know what to refer to Katie as, was she an RV, coach, camper, motorhome, or even a bus?

'What are you doing?' Gail asked.

'I'm thinking.'

'You're always thinking.'

'Well, actually I was on the Internet looking at a forum for RV owners in the States.'

'What do they say?'

'They had a special section where newbies owned up to the mistakes they made on their first trip, it's a long list but really funny.'

'Like what?'

'One chap was filling his toilet tank half full with clean water to rinse it. He got chatting to a passer-by and forgot how long the water had been running because after a while is started "raining". Unfortunately the "rain" wasn't coming from a cloud; it was coming out of his black tank vent pipe on the roof. A lot forgot how high the RV was and took tree branches with them, others left the aerial up or the door steps out and drove off – one even had the electric cable still hooked up. And here's the funniest – a wife had been sleeping on the bed and got out at a truck stop to stretch her legs and use the restroom while her husband was paying for fuel. The husband returned and set off thinking she was still in bed asleep and didn't notice she wasn't until 60 miles down the highway.'

'I hope were not going to embarrass ourselves with anything like that. Don't you think we should have a practice first? After all you've not driven Katie, and we don't even know if everything works or how to work it?' Captain Sensible suggested.

'Well Nigel said he would show us, but I know what you mean; if something goes wrong in France we could be in trouble. I'll give it some thought.'

'You're always thinking.'

And later: 'I might have the answer. We could take Katie for a weekend to the Caravan Club site at Abbeywood.'

'It sounds nice, where is it?'

'Thamesmead.'

'You mean *The Thamesead*, the housing estate near Woolwich, that's not very far or appealing?'

'Well I agree it's only twenty-seven miles from Nigel's but more importantly only nine miles from the bed in our flat.'

'You feel confident then?'

I don't know what I felt. I only wanted to be off and the whole thing seemed to have dragged on for an eternity with all the legalities and driving training and tests and another to come so I was at a stage of almost being tired of it. Then perhaps there was a fear of failure and if you never start you can't fail but then again it was about to get exciting.

On Saturday May 31, the weather forecast showed beautiful golden suns all over the map of the southeast, so I wore my shorts in anticipation. The journey to Basildon took 45 minutes from Limehouse station, on the C2C rail service.

We sat opposite three large and noisy Essex girls (well they were travelling that way). They wore skimpy clubbing clothes that did little in the beauty enhancement area but large legs in micro skirts (well not much was in them) still held a curiosity in their inappropriateness. They had obviously been out all night and whereas one couldn't stay awake the other two were high and lively, laughing out loud. Perhaps they were discussing the blokes they had met or, were they sniggering at the older bloke opposite and his inappropriate shorts that early in the morning?

We all descended at Basildon station (see, I knew they were Essex girls) but we took the only available taxi and were at East Coast Leisure in less than ten minutes. Nigel's own magnificent blue and cream 39-foot Monaco Diplomat was parked outside and looked enormous. Would I cope with Katie? She was inside, clean and sparkling and also looked huge. Nigel had been up late washing her, polishing those wonderful shiny 19.5-inch stainless steel wheels and blacking the tyres. She had been fitted with solar panel, satellite dish and satellite navigation (satnav) system. Boris, who we had delivered earlier in the week, was at the rear of the workshop, the A-frame attached. It appeared complicated and it was a relief to know we were not going to use it that day.

Based on my reading of *'The Complete Idiot's Guide to RV'ing'* I had prepared a questionnaire that started with 'Arriving at the site and setting up' followed by 'Living in the motorhome'. If Nigel could take us through that, we would be able to set up and use the facilities without making complete idiots of ourselves. My greatest fear was doing something that drew attention to us or, worse, damaged Katie. By going through the list we would also establish that everything was in working order. As we followed Nigel round, pressing, opening and closing, pulling out, switching on and off this and that making hurried notes I realised just how much was in Katie and how complicated she was. My head ached with information and 1.00 p.m. was upon us. To stave off the hypoglycaemia we called time and Gail and I went to the local Burger King.

As I sat down with my Filet-o-Fish meal I wondered aloud,

'Do you think we can remember all of that?'

'I'm sure we will and you made lots of notes.'

'The toilet tank emptying seemed critical and potentially embarrassing should it go wrong. Are my notes going to be enough?'

We wandered back. It was a beautiful hot sunny day, and the choice of the shorts fully warranted. While Gail explored the shop for everything apparently indispensable to the modern motorhomer Katie was moved out of the workshop and Nigel explained how to deploy the satellite dish, regular TV aerial, the awning and slide-out. How embarrassing would it be to get the satellite dish stuck up or the slide-out out? Finally we discussed the route to Abbeywood. But why were we doing this when we had a new and expensive satellite navigation system installed? Nigel sat in Katie and programmed it. Looking over his shoulder, my confidence waned as I saw *'Piccadilly Circus'* (twice) as a destination as well as *'Basildon Hospital'*. It was time to say thanks and goodbye to Nigel and wife Lynn and set off on our epic 27-mile adventure.

As I sat behind the wheel it crossed my mind that I was about set off in a vehicle I had bought that I had never driven, or been driven in, and I would be driving this left-hand drive thirty-three-foot motorhome after four days training on a right-hand drive box van along a route I didn't know. As we had decided not to tow our car Gail followed in Boris and we communicated with walkie-talkies we had bought.

I drove down the road 'In two-hundred yards turn right', it was the satellite navigation, followed by 'In three-hundred yards turn left then there is a roundabout take the second exit.' I had a map to follow on the screen but I was concentrating too intensively to look.

'You're running over the centre white line' my walkie-talkie crackled.

Getting used to the left-hand drive strangely meant I didn't run close enough to the nearside. I adjusted the mirror and things improved. I went down the A127, with beautifully timed instructions from the satellite navigation then onto the M25 and speed increased rapidly up to 55mph without trying. Katie was delivering a lot of power. I even overtook a truck. What a feeling being up at the same height as the truck driver and as he was right-hand drive, I was right next to him. Should I wave or not? What would be customary? No, best to keep my hands on the wheel.

As I approached the QE2 Bridge Katie automatically kicked down a gear and climbed the slope then powered over the top and down to the tollbooths where I would be on the wrong side. A quick squawk over the Walkie Talkie and Gail drove past and up to the booth to pay for both of us. The entrance looked narrow but if those continental trucks could get through then why couldn't I, after all they were, oh no, the same maximum width! I crawled cautiously through.

As we left the tollbooth I settled back confident of a straight motorway run to the A2 when suddenly that voice intruded, 'In five-hundred yards leave the motorway by the exit road.' Like an automaton I obeyed and let Gail know this was unexpected but 'I have been told to do it.' What were we going to get into, little lanes, narrow high streets, or, low bridges? The instructions came thick and fast but very clear. We did have narrow streets but all were negotiated and before long we were at a roundabout and I could see the *'Camping and Caravan'* signpost. We went up the hill, turned sharp right then left into the campground 'You have arrived at your destination.'

The warden was helpful. We explained we were camping virgins and he pointed out a large pitch for us so that we wouldn't collide with or set fire to anyone else. Ian the assistant warden then set off ahead on his moped to warn everybody and helped with the reversing. The campground was an oasis in an urban concrete jungle with lots of open space surrounded by trees all kept immaculately like the grounds of a geriatrics' nursing home. Left alone, but I sensed with all eyes watching us, I went inside so that no one would see me reading the instructions from the morning's induction; they looked inadequate against what we had to do.

I started confidently and methodically plugging Katie's electrics into the pitch supply post then neatly put away the orange cable under the motorhome so that no one would trip over it.

Before we could deploy the slide out we had to level Katie using the joystick controls and lights on the hydraulic control panel on the floor behind the driver's seat and as a secondary check the swinging of the fridge door. First the rear then the front two shiny jacks descended from Katie's chassis and took the weight, this being confirmed by four red lights on the panel. I moved the joystick in the direction of the amber light and Katie rolled side to side and pitched back to front until it was extinguished.

It was 30 degrees Centigrade and we were sweating so we decided to try the air conditioning but no, nothing, zilch, breathless. In the lounge we re-checked the settings on the wall control panel. This had been difficult to understand even when we went through it with Nigel. It was there in block capitals 'COOL, 65 degrees', and fan on 'HIGH'. The perspiration increased.

We checked a socket with a hairdryer – no power. At this stage and with the heat beating down, we tried to think logically but didn't. Could the camp's supply provide enough power to drive the air conditioning? We tried the generator, knowing this would. I pressed the start button on the dashboard and it sputtered into life. A downdraught of cool air whooshed its way into Katie but we couldn't run the generator all the time, as it was too noisy.

We couldn't think of any logical reason why the 16-amp supply I'd plugged into shouldn't have provided the power. We were going to be limited without electricity. Had we blown a fuse? We checked the two main 'house' fuse boxes in Katie's bedroom rear overhead cabinets and found one switch was down while all the others were up. Referral to the checklist showed this was how they should be 'Don't touch the right-hand one' Nigel had warned. Back to the beginning, the only way power could get in was that orange cable. Gail checked the electric connections and discovered the notice on the camp supply post about turning the plug to the right to lock it in position – idiot me! We now had power surging forcefully through the orange cable, which I now saw, had been dangerously close to being squashed by one of the jacks.

We now returned to the list and went into the bathroom. For fun we checked the levels on the computer panel above the sink and the lights glowed in a dazzling diagonal display. We were nearly full with the fresh water and LPG, the battery power was at the maximum and we were empty on the grey (shower water) and holding or black (toilet) tank. We turned on the water pump, the switch glowed red on the panel. Next we turned on the hot water taps, and initiated a flow so there were no air locks and we had water in the water heater. We turned on the electric water heater and the switch glowed a comforting neon wobbly red.

We checked the large American fridge and freezer and set the switch to auto so it would use electric when available or LPG when not, then we set the temperature and produced that comforting green light. We checked the gas cooker rings and oven and they all lit up straight away.

Getting increasingly confident, we pressed the slide-out button momentarily then stopped – was the driver's chair forward, had we removed the holding stays, were the jacks down, and was the foot brake on? The slide out groaned, creaked, and shuddered and was then on its gravity defying way creating our spacious home. We decided not to deploy the large outside awning, which might have been riding our luck too far.

We now tried all the lighting combinations, how atmospheric it was going to be in the evenings. Was the microwave functional – it was flashing 'SHARP SIMPLY THE BEST'? How will we be entertained? The large TV dominated the lounge. I raised the regular aerial, but couldn't rotate it and went outside to have a look. Yes it was up. I came back inside and tried again but it wouldn't budge. We decided we might break it, so we left it alone. The TV was giving a good impression of a winter sports snowstorm with the possibility of ITV just coming through but hardly watchable.

I switched the satellite on and produced a humming sound, then clicking and more clicking. I went outside. The dish was doing an excellent interpretation of a ballerina pirouetting round and round, then a pliè up and down. I guessed that the trees were blocking reception. We had the luxury of two televisions (more than at home) but no pictures.

I tried the radio that played in the lounge and bedroom to find it picked up every rap station in South East London, including the one that kept selecting itself on our home radio and was therefore our most hated. No BBC, only rap after rap after rap. I went to the car for a calming Diana Ross tape. Then a discovery – you could turn the radio on and off from the bedroom, what joy.

We had so much more to check and discover it was exciting but the fear always lurked that something might not work.

The lounge sofa bed pulled out to reveal more storage space. The dinette table removed; the dinette seats made into a bed and had storage drawers underneath and lifting the seats out showed that the drawers didn't go all the way to the back and we had more storage beyond the drawers.

The water was now hot and the commodious shower and all taps worked. The toilet filled with water from the right-hand pedal, then when the left-hand pedal was pressed disappeared with a satisfying throaty gurgle down the hole with a rinse of fresh water. The little flexible showerhead attached to the toilet could be used to rinse the bowl of any residual nasties. The roof fans worked and the 'magic' fan in the lounge blew in or out at different speeds. We found and tested the alarm systems; the smoke alarm needed a new battery.

We unloaded our stuff from the car. Katie had started to look like a home –we had bath towels in the shower room, sheets on the bed, food in the cupboards, all essentials not a knick-knack in sight. I played with the satellite one more time and it locked on to Astra 2 and all sorts of Sky channels were beaming into our cosy motorhome. We ate and went to bed exhausted but it was still hot and we didn't want to run the air conditioning through the night. It was hard to sleep. Breakfast was a pleasure sitting at our dinette table looking out onto the grassy campsite and its many trees.

Like a coach, Katie has 'belly' lockers along her full length and many join side-to-side giving a large carrying capacity. From these we retrieved loads of stuff that had been put into the motorhome by the first owner but looked unused. Everything was washed and cleaned. We tried our new vacuum cleaner purchased from John Lewis especially as a balance between weight, sucking power and electric consumption.

Then we had visitors. Trevor from East Coast Leisure arrived with his wife Avril to see how we were getting on. Did they visit everyone who bought a motorhome?

Apparently Trevor and Avril park their own motorhome at Abbeywood, so they were regulars and not far away. A welcome visit nonetheless. Coffee was made and we spent hours talking about our experiences so far. Trevor couldn't get the aerial to rotate either and it will need freeing off.

At night we began to enjoy the feeling that everything was under control, the temperature was cooler and we looked forward to enjoying a well-earned rest.

We dropped off to sleep with no problem but were soon woken by rolls of thunder, flashes of lightning and the drumming of lashing rain on the roof. Immediately the adrenalin was coursing round prompting all sorts of questions. Gail thought the satellite dish was going to attract the lightning and we would end up fried to a frazzle. I struggled to the front to retract the dish then went back to bed. No sooner had I done this than Boris's alarm went off wailing like a banshee on heat. What an embarrassment, our second night on the campsite and we were waking everybody. I donned a coat and shoes and stepped out into the pouring rain.

As I approached Boris and was about to fire the remote his lights flashed and the alarm stopped. What a mystery. I could understand the thunder setting him off but switching it off as well? Drenched, I retreated to Katie. Gail was in the doorway laughing – with Boris not far away she had fired her remote through Katie's windscreen. I towelled myself down and went back to bed.

'Yes, very funny, but can we get some sleep now?'

The next day we rose slowly and decided to empty the toilet and shower waste tanks (after two days camping, two 35-gallon waste tanks and no diarrhoea). We drove down to the campsite dump station and everything went like clockwork. The snaky hose was connected to Katie's waste outlet then dropped down the drain. Gail pulled the holding tank lever and down went what few digested brown trout were in there. She then closed the lever and released the grey water and satisfying gurgling sounds were heard as the shower water disappeared down the hose taking any 'nasties' from the holding tank with it. We cleaned up. We then put some water in each and added what we thought of as a reasonable amount of the appropriate sanitizers to control any smells. With that we left.

We arrived triumphantly back at Katie's parking compound and with consummate skill squeezed her in. We had thoroughly enjoyed our limited experience. It was over in a blur.

Bigger Yellow Truck

On Monday June 9 I struggled to get all twelve feet of Boris cleanly parked against the kerb, yet in twenty minutes I would be driving and parking nearly forty feet of truck.

At 1 p.m. Roger, the instructor and company boss, arrived from his previous lesson and we clambered aboard the bigger yellow truck. Roger explained the more complicated eight forward driving gears and a crawler gear on a 'splitter' box, what gear to start in when going up a hill or down a hill and when to skip gears as though it was an everyday event. (Well to him it was.)

'Please drive on!' he barked.

I started correctly using all mirrors, checking the blind spot and signalling. The truck hardly moved in second or fourth gear and I had trouble getting the higher gears. Roger waved his hand in a get-on-with-it kind of motion.

'Today is a honeymoon period to get used to the gears and positioning the vehicle,' Roger explained as I missed sixth gear again. After four hours I had driven the truck through some of the narrowest streets in Enfield, Chigwell, Waltham Forest and goodness knows where else and negotiated some vicious left and right turns. At the end Roger drew the gears out on a piece of paper for me to take home like a naughty schoolboy. I knew where they were, damn it, it's only when I approached a hazard and had to change quickly that I forgot. At home I had aching arm and shoulder muscles. Gail found a tube of *Deep Heat* we hadn't used for years and I applied it too liberally. I felt as if I was in flames.

On day two Roger confessed that after Monday's performance he was deeply worried. By Wednesday I managed the gears much better but reversing was another matter. We practiced on a Leisure Centre car park and the right-hand downs and left-hand elbows up when you see the trailer in your mirror all became blurred together and I lost all sense of what to do to get the trailer moving in the right direction or, correcting it when it became wayward and off-course. Roger despairingly got out of the cab and walked off to make business calls on his mobile in the belief that left alone it would come to me.

Things were so bad that on the way home I stopped at Tesco and bought a toy articulated truck. Unfortunately the toy's front wheels didn't turn so it wasn't possible to mimic the exercise. I wrote out on a sheet of paper what I should do at each point on the course and hoped this would help.

We tried uncoupling and re-coupling the trailer.

'Put the trailer brake on and then wind down the legs.'

It was harder physical work than I had imagined.

'Detach the electric cables and air lines. Don't forget to remove the number plate then detach the dog clip and pull the large handle out, forward and out again to release the king-pin and the trailer.'

It was stuck in. I got back in the cab and drove forwards then backwards to try and release it. It still didn't come. Roger in frustration pulled it out while I was in the cab perhaps thinking I was never going to manage it. I drove the tractor forward and the trailer dropped onto its legs. Coupling the trailer back on required you to reverse everything and act like a drama queen and shout out what you were doing. Reversing the tractor onto the trailer was a lot easier than getting it off had been.

The next day the reversing was better and I did the uncoupling without reference to the notes but I still couldn't get that trailer freed up.

Friday June 13, was an auspicious date for my test, set for 1.00 p.m. In the night I had rehearsed my uncoupling and reversing procedures over and over in my mind. Because two of us were taking the test in the same truck that day we had to share our time. Dean would drive from 7.00 to 8.00 a.m. and I would arrive at 8.00 a.m. for an hour's drive then Dean would have an extra hour before his test at 10.15 a.m.

I didn't drive as well as I would have liked and realised I was not at my best in the morning after a poor night's sleep. Roger wasn't doing so well either getting a text message on his phone that a potential business partner would not be buying in after all. On the other hand young Dean drove like a professional with smooth gear changes, good anticipation and at speed but within the limits. Roger had often told me to 'push on.'

We went to the Leisure Centre car park and set up for reversing. I went first and somewhat to my surprise managed to reverse satisfactorily three consecutive times. Dean then did the same. As the other truck arrived for their pre-test reversing practice and Dean realised he had an audience he did the manoeuvre slightly wrong then tried unsuccessfully to get out of it by some exaggerated movements and made it worse.

We drove to the test centre where Roger gave last-minute instructions like a general before the battle:

'I've no problem with you, Dean, provided you don't get over-confident,' he pronounced obviously expecting a pass.

'Keith, you must watch your corners and not be too cautious at roundabouts when they are clear,' he added, almost as an afterthought.

'We've got one of the best pass rates here at fifty per cent against the National average of forty per cent so we should be OK today.'

Excuse me, wait a mo, hold your horses, I thought. If he's expecting Dean to pass, where did that leave me – probably in the pay for more lessons; you look like you can easily afford it and we need the money category. Dean magnanimously confessed to some nerves and that he hadn't eaten properly I offered a muesli bar that my thoughtful Gail had provided for me.

At the test centre I tried to relax by sitting on one of the white plastic chairs in the scrubby garden area and mentally detaching myself from the comings and goings associated with a number of vehicles, their drivers, instructors and examiners. I saw Dean drive out of the centre in the truck after completing the reversing and stopping tests. Roger and I went for a meal at the Black Cat transport café. Roger had the *'Jubilee'* breakfast with absolutely everything and I the ham, egg, and chips.

When we returned, our truck was already in the test centre with Dean and the examiner sitting in the cab. Roger wandered over and they all came back together. Dean had only incurred four faults out of an allowable fifteen for his on-road driving but had failed the mandatory reversing. General Roger was visibly shaken, thinking he might have a double whammy on his hands and along with the setback to his business plans, Friday 13th was going to live up to its name.

We set off on my final hour's drive. Roger was not in any mood for conversation. At the test centre we parked up and I inspected the toilets. Gordon was my examiner and he seemed kindly enough. We walked onto the manoeuvring area and he explained the reversing requirements, I double-checked the number of 'shunts' allowed. He would allow two, provided they were needed and the vehicle was under control. I climbed in the truck and drove it to the correct place to begin the exercise.

Reversing, so right hand down, all-round observation, follow the yellow line with the front wheels and check when trailer is straight. What the hell had happened? The trailer was already coming back to the right towards cone B. I couldn't have straightened my wheels early enough. What should I do? If I hit cone B that would have been an end to it, if I took a shunt I only had one left yet I was still only just starting the exercise so surely I would need that shunt later? I tried a severe turn up with my left elbow but the trailer was not responding fast enough and the cone was too close now.

I took a shunt. My memory is hazy about the rest. I just remember the trailer going diagonally across the yard. I stopped and took another shunt to make sure the trailer was close to straight and I was in line with the bay. That was it, no more chances. I told myself you have to take this back in a perfect straight line with no room for error, otherwise it was all over. You could do it if you remembered 'if it appears in either mirror then turn the steering wheel in that direction.'

That final line never seemed to come but I knew I mustn't go faster because the probability for error increased. I was finally in position. Gordon climbed into the cab.

'All right, Keith, we are now going onto what is known as the stopping test. I want you to drive...' It went perfectly.

Out on the road Gordon and I chatted about motorhomes, truck driving as a profession and road planners not giving due regard to the difficulties they created for trucks. He demonstrated this by asking me to drive around an acute left-hand bend with railings on the inside and bollards in the middle I remembered Roger's caution about my rear wheels and pinched some road on the way in and out of the corner.

As we came down through a shopping high street someone had double-parked their car on my nearside facing the wrong way and immediately opposite some central bollards. Was there room? I started to crawl past with no more than an inch on each side. Gordon looked out of the cab window then spotted the anxious owner in the shop doorway and shouted, 'I'm glad it's not my car.'

Was this good or bad news, I wondered? I glanced at the clock in the middle of the speedometer; things were going reasonably but the longer it went on the greater the chances were that someone could force me into an error. Keep the speed going I thought this is a forty-mile-an-hour stretch, up to seventh gear and then the roundabout known as Six Flags appeared. This has a raised centre with particularly lush vegetation that actually prevents any view of traffic coming round from the far side of the island even from the height of a truck cab. A truck came round and was going straight on thereby blocking traffic from the nearest entry road, I shifted down to fifth gear and we went through smoothly.

Back at the centre it was trailer time. Roger had wandered up and was already talking to Gordon. I commenced the exercise, acutely aware of both of them watching closely.

I put the brakes on the trailer and started to wind down the legs. I was already sweating from the driving test and in the mid-afternoon heat the physical effort of getting those legs down became a challenge. In my imagination I sensed an air of disdain from the observers as this City guy sought to enter their hard-bitten world of the trucker. Then horror of horrors, having got the legs down, that bloody handle wouldn't come out and release the trailer king-pin. Of course I had never released the trailer king-pin because Roger seeing me as a ten-stone weakling had always stepped in at the last minute to do it for me. I climbed back in the cab and joggled it back and forward (it should be backwards) then climbed down wondering if this was going to work for me. I stared at the lever, then like an Olympic athlete visualised it coming out as easy as a knife from butter and pulled that lever out, forwards, and out again as though I was Arnold Schwarzenegger and nearly fell on my arse in surprise. I completed the exercise.

Gordon climbed back in the cab.

'Do you mind me discussing your results in front of your trainer?'

'I'm pleased to tell you that you have passed with eight general faults, none of which were duplicated. Congratulations.'

General Roger's record was intact.

I called home to tell Gail to break out the Yorkie bars again.

The Journey Start – France

It was Monday September 22, and Katie once again imposingly occupied the taller half of Nigel's workshop, she looked much bigger in the confined space towering over us. The generator, that had gone into a sulk and refused to start, was transplanted with a long-awaited circuit board from the USA, and roared into life like a dozing lion prodded with a hot iron. We had no more excuses and were free to start our European tour and new lifestyle. I was excited and anxious all at the same time wondering what challenges awaited us?

I drove Katie back to her parking compound stopping en route at the BP station to fill her half- empty tank with £150's worth of petrol – ouch, reality hit home! We also topped up the 27.4-gallon LPG tank for the first time. It is only used for room and optional water heating, cooking, and the fridge will switch to gas if no electric supply is available. Katie's LPG tank is behind a locker door about midway along the passenger (right) side. The filling cap was removed, a brass adaptor fitted, the pump nozzle placed in, the handle locked, the button on the pump pressed continuously, the tank topped up then the procedure reversed with a little whoosh and smell of gas at the end.

I offered the same credit card used for the fuel for the £9.11's worth of LPG – it was rejected. The young and spotty assistant in his oversized green Bolshy Person uniform asked me to wait while he went into the back office, my credit card disappearing with him. No one else was serving. I occupied myself staring at the tempting array of sweets and chocolates wondering whether to have a trucker's Yorkie bar or a Twix. Coughs, sighs and shufflings of silent frustration and exasperation that we British are so good at, emanated from the queue. The short, rotund, dyed-haired, lady manager eventually appeared on the elevated platform behind the counter and with a great sense of self-importance explained to me, and the lengthening queue, that there was an alert on our card. How could that be, I pleaded, I had just paid for £150's worth of petrol with it and now all I wanted to pay was a measly £9.11 for the LPG; I am Mr Honesty himself. The hungry, sandwich and soft drink clutching lunchtime populace, many of whom looked like they were appearing in double or drop, was having none of it and had already arrived at a guilty verdict and was only prevented from giving a thumbs-down by the thought of dropping their sandwiches and losing the competition. I paid with a debit card, hung my head in shame, and left ignominiously.

'How are we going to manage without valid credit cards for our journey and how can replacements be sent if we have left the country?' I asked Gail.

'Can you tell me your mother's maiden name and your last and rather extravagant purchase? Our computer had you down in the Mr Miserly category then you suddenly decided to become Mr Profligate without the necessary authorizations are you surprised we pulled your card?'

Didn't they realise we now owned an American RV that had to be filled with wonderfully expensive gadgets and that was just half a tank of fuel wait until we did a whole one? I wanted to respond more belligerently for the embarrassment they had caused and their faceless pompous attitude but we needed the cards so acquiesced in a grovelling manner.

We drove Katie back to her parking place and returned home running through a torrent of rain. We had spent two days lovingly polishing Katie and Boris.

The day of departure for Folkestone arrived. Everything readied in the spare bedroom at our flat was packed into suitcases, rucksacks, and plastic bags then carried down to the garage and squeezed into Boris with his rear seats folded down. Gail gave our flat a final clean and would probably clean it again immediately on our return (why do women need to do this – who will know?). Believing that all food would run out beyond the Channel we made a final shopping trip to the local Asda and then were on our way to Essex. It was a lovely day with clear blue skies but hard to believe that the next day we would be driving Katie towing Boris through France. Shouldn't we be working?

When we arrived at the parking compound, Gail stowed the stuff from Boris into Katie and I tried to make Katie presentable by removing the purple pigeon droppings that had spattered the sides and windows.

We hitched Boris onto Katie with the A-frame for the first time then had to drive out of the yard avoiding the old roofing tiles, bricks and chimneys that probably had an historic preservation order on them stacked each side of the narrow roadway. No problem for a licence carrying truckie, I thought.

'Can you reverse?' – Gail was walking behind and on the walkie-talkie – 'Your rear offside wheels are about to crush some tiles.'

'No!'

She had forgotten that with Boris in tow I couldn't reverse. Gail squeezed between the tiles and Katie and removed them. At this point I became conscious of the standing crowd at the windows of the adjacent offices. We were at the start of our tour of Europe and were struggling with an obstacle twenty yards from the start with a bloody audience – how embarrassing.

Katie pulled Boris, a forty-eight-foot-long combination, onto the A132.

We had a smooth drive down to Black Horse Farm Caravan Club site near Folkestone and reversed into one of their good-sized, rectangular, stone-chipped pitches cut into the grass getting settled like professionals. With the slide out we flopped into the chairs in Katie ready to enjoy a cup of tea, hardly believing we were on the way, but we had forgotten she had no water in her tank. I drove Katie out of the site, turned in the car park and drove back in so I could reverse into the tight pull-in where the water supply was located, overlooking that two of our hoses linked together and run through a hedge would have reached the tap from where we were originally parked. Finally, we were settled and enjoyed that cuppa, then dinner with a previous colleague who lived locally and had been brave enough to buy my business and still wanted to speak to us.

'Don't you miss the office and city life?' he asked.

'I wouldn't mind the money' – but I couldn't wait for the morning – it seemed unreal, like we were cheating in some way.

As the morning dampness evaporated, we hitched Boris up smoothly in the campground and drove down to Dover docks. We were on full alert for signs to P&O. When obstacles approached Gail rose out of her seat, leant on the huge, flat cream mock leather dashboard in front and peered through the massive windscreen. The car ahead stalled at traffic lights and wouldn't restart. Fortunately, and remembering our reversing problem in Wickford, we were far enough back and could pull round.

The ferries came into sight – what lane should we go in to show our tickets? Were we a truck, a coach or, a car? Thinking this would not be a good place to make a mistake, I stopped well before the long line of ticket booths spread out across the tarmac like some East German border post. Gail jumped out and ran to ask at the nearest kiosk.

'Any one will do.'

We drove into a lane where the attendant was on the wrong side and too low for Gail to reach down from our high cab so she had to get out and walk round to the booth.

We were early and offered the prior ferry. We queued in lane 65 flanked by articulated trucks as long as us and behind some smaller European motorhomes and pretended we had done it all many times before and hoped we wouldn't make some ghastly blunder.

'I think I'll put the French satnav CD in to save us panicking when we get off the ferry in Calais.'

'CD not recognised.'

'Well we're not in France yet.'

'It's only twenty-two miles; if it wasn't misty I could see it.'

Looking in the rear-view camera I saw three swarthy East Europeans curiously examining Boris and the A-frame.

'Where are you going?'

'There's a gang of East Europeans around Boris. They may be trying to stow away in him.'

'Why would they want to do that? It's illegal *immigrants* that are the problem, not *emigrants*.'

'They've probably been here for a while and found out what it's really like under Blair's government.'

I descended to find out they were curious as to how the A-frame worked so I explained the technology.

'It's simple. When I drive forward, because the car is attached by two rigid steel bars it tends to go forward as well.'

'The steering is locked?'

'Oh no, it is free and turns the corner as we do, occasionally my wife rides in the car if we expect lots of corners,' I teased.

'Amazing – and the cable?'

'That is for the brake. When I press my pedal in the motorhome the car brakes go on as well.'

'He says a cable comes all the way from his motorhome brake pedal to the one in the car.'

I left them with this improbable thought as we were called and drove up the ramp onto the boat deck and down the extreme left-hand lane that appeared too narrow to accommodate us and worryingly swept in as we approached the bow of the ship. People descending from tour coaches and getting out of cars pointed us out to their neighbours and curiously looked at Katie, Boris and the A-frame.

The channel crossing was smooth and 1 hour 20 minutes later and proud purchasers of many cut-price videos we were back in Katie. We exited after most of the trucks, looking carefully for the widest lane on the dockside and chose the coach route but they all seemed to end up at the same place. Before we knew it, and not realising quite how, we were on the A26 motorway. What a beautifully clear, blue-sky day. Where was everyone – had we unknowingly caused an incident stopping all traffic behind us? The smooth tarmac motorway was quieter after the congestion of Dover and the flat agricultural countryside effortlessly rolled by.

'Try the satnav CD again.'

'CD not recognised.'

'I should have known. I think I've bought a dodgy CD. "France" scrawled on it in marker pen and half normal price should have been a clue.'

'So we've paid twelve-hundred pounds for a satnav system to guide us eighty miles from Wickford to Dover?'

'No, only six-hundred; we'll be able to find our way back.'

'*Péage* (toll) ahead.'

'Which lane?'

The gaps looked surprisingly narrow for our eight-feet-four-inch width and there were plenty of scrapes and paint marks on the massive concrete blocks guarding the entrances.

'Look, there's one with a Union Jack on it.'

How thoughtful of the French to have a lane just for the British – what *Entente Cordiale*. But it was a mistake for a left-hand drive vehicle so Gail had to take the ticket from the automatic machine that was now on her side.

The road remained quiet, the sun shone and 55mph was comfortable. I tried to enjoy it, sitting high in the cab, a panoramic vista in front, feeling in control, even relaxed, but still mentally going through events so far that day constantly checking the mirrors for road position and the rear-view camera to make sure Boris was still there.

'Did you put Katie's number plate on Boris?'

'No, did you?'

We had been illegal since Folkestone. We were going to have to pull off into an *aire de service*, what would it be like, were we going to have difficulty?

It was neat and clean with a good series of drive-through slots for cars and trucks and plantings of trees and bushes. I found the number plate in the locker and, relieved, took a photo to record the moment. Gail rather than being relieved was probably wondering what else I had forgotten or could go wrong.

The drive was steady and after 60 miles we pulled into another *aire de service* and parked with the trucks – it was tight. I checked around. The brake lights on Katie and Boris were on and remained on whatever I did. A truck pulled in alongside with a couple of inches to spare. I smiled at the driver as a fellow truckie and indicated the tight space with my hands but he was uninterested and set off for the bogs. I wondered whether I could get back into Katie through the side entrance door as he was so close but did.

We sat in Katie and ate our sandwiches quickly. It was annoying that the brake lights stayed on and I thought this would drain Boris's battery as we drove along and knew I would worry about it. (*This is not, of course, true as power is taken from Katie and the reason the lights were on was because we were parked on a down slope and Boris was pushing against Katie activating the brakes.*)

We had pre-booked our first night in France at a campground near St-Quentin and directly along the A26.

We turned off at Junction 11 to Essigny-le-Grand and drove along the D1 into a village of cream-coloured terraced houses and shops dirtied by ages of passing traffic but still charmingly French and took a right turn. I visualised the map from viewing it at home on the website. After a narrow section and some more houses we left Essigny-le-Grand behind, going towards Seraucourt-le-Grand along the narrow D72 raised above the open flat ploughed fields like a fenland road, my nearside wheels brushing the grass verge (and to Gail's mind too close to the ditch) a narrow road for sure. What's that in the mirror? A huge French articulated truck was rampaging and snorting behind us. Suddenly I was Denis Weaver in the Spielberg movie *Duel* sweat pouring from my brow then it became too real as he forced his way past us at speed – crazy – there can't be room? What a heart-stopping moment – he definitely won that one but we were shaken and slowly drove on to Seraucourt-le-Grand.

After a couple of turns in the village with its pretty stone houses we arrived at the entrance to Camping Le Vivier aux Carpes. The gated entrance, in what appeared to be no more than someone's driveway in a suburban street, was narrow but negotiable. Gail descended and dealt with the formalities in the office. The lady then jumped on her sit-up bicycle and led the way as we drove past the reception block and caravans to one end of the site where we were given a grassed area to ourselves with the river running alongside and ducks and geese everywhere. She explained the electricity was *'sur le mur'* of the barn behind us. I swung Katie round in a full circle and faced the river. We decided to keep Boris hooked on and set Katie up – she was virtually level but we still put the jacks down and extended the slide-out.

We sat outside in the glorious afternoon sunshine enjoying a cup of tea, phoned our respective parents to let them know all was well, and glowed with a sense of achievement (we had driven all of 117 miles from Calais, around 200 miles in total).

Gail made friends feeding one of the ducks with moistened crusts and I busied myself taking a photo. We enjoyed our well-deserved meal, watched TV – it seemed an age since we left the Folkestone campground and what with ferry kiosks, missing number plates, brake lights staying on and a mad French truck driver, the first day had been a full one not without its tension. We had our individual snoozes.

We had a restless night. I was in a struggle for survival with a blue-overalled, beret wearing trucker progressively squeezing past Katie on a road that became narrower and narrower until I had to resort to driving across ploughed fields Boris bouncing along behind. Eventually I escaped to a duck-filled watery meadow and come morning Boris and Katie were completely covered in droppings from ducks sitting and squitting on the roof. In the morning the ducks and moorhens were all around but Katie and Boris were unscathed apart from the windscreen and front that were covered in sacrificial offerings in brilliant hues from yesterday's kamikaze flies and other winged insects.

As we didn't have to unhitch Boris and put 'LONG VEHICLE' signs in the back window, things were much easier and we were soon ready to go. We steadily drove out of the campsite, saying to each other how delightful it was for our first experience in France then I suddenly remembered to drive on the right. We made our way through Seraucourt-le-Grand along the D72 back to Essigny-le-

Grand a quick left and right and under the tunnel (we had checked the height on our Michelin map), to reach the north bound D1. This was an impressive road, at least as good as the toll roads we had been on. Soon we were back on the A26 motorway and travelling south.

We had made no bookings for that or any other night and were undecided where to aim for – was this sense of bravado and pioneering spirit right I wondered.

The countryside was flat and uninteresting, the road quiet particularly compared to a British motorway. We cruised along at a steady 55mph and almost relaxed.

Even though the tank only showed half empty we stopped at a *'TOTAL'* petrol station to fill up and have a break. Access was good. As I opened the filler door I came to the realisation that petrol caps do not automatically reinstall after refuelling and ours must still be at the 'BP' station at Wickford because of the diversionary fracas and fury about the credit card. Gail went to pay the bill by credit card it was 132 euros and the till girl was a little taken aback.

'Trés, trés, grande' – did she mean the bill or Katie?

I chatted to an English lady of middle years in a Smart car. She says she has only just made it to the garage – 'running on vapour' – I thought they did anyway. I bought an emergency petrol cap in the shop.

Reims provided a little more traffic and interest.

After 120 miles we pulled into Aire Champ-Carreau, for lunch using the lorry park. The French motorway services were increasingly impressive with plenty of space and good pull-through sections for caravans as well as trucks however, the toilets had crouchers so I was glad I had only gone to stretch my legs and urinate.

With more hills Katie automatically dropped into third gear and picked up her revs to maintain speed. Truck drivers were courteous as we passed or were passed. Little caught our attention. A coach cruised past and on the command of the tour guide, who also hadn't had much to point out for a while, every one of the Japanese tourists from both sides craned over in perfect synchronicity, and a red carpet, Hollywood style, volley of camera flashguns erupted.

Péages provided exciting moments; we thought we were getting good at squeezing through, Gail standing up out of her seat to double check me, but we watched large trucks taking them at speed presumably with an inbuilt token device. I was always ready to shove a large Euros note at the cashier and let them do the work of counting out the change, particularly as we never knew what rate they would charge us – motorhome, motorhome plus trailer, or truck. Being the driver and not the chief cashier however, I didn't have that opportunity. Because Gail didn't want to carry round a heavy purse she would meticulously count out onto her lap a mountain of small denomination Euro coins as though coinage was going to be cancelled in the next hour and then I would try to collect it all in my hand and attempt to get it across that gulf between the driver's window and the safety of the extended cashier's hand way below. Because my window was of the sliding type, thereby cutting the opening in half, I was reluctant to squeeze my head out as some childhood experience suggested that once my ears were through and opened out going back was not an option. The time already elapsed plus the time for the cashier to count the change all over again provided ample opportunity for those queuing behind to examine every detail of Katie and Boris so they didn't need to come and ask us irrelevant questions although the look on their faces in the mirror did suggest they might have some.

We discussed where to stay, and consulted the *Caravan Club, Caravan Europe* – should it be Langres itself or Humes-Jorquenay? The latter seemed easier to negotiate but we would see nothing of Langres and I had read it was supposed to be an attractive walled town. We pulled into Aire du Bois Moyen and decided to go for Humes.

Beyond Troyes the A26 became the A5, then eventually signs appeared for Langres Nord and we had a short drive along the A31 before coming off at the N19 exit *péage*. After this we joined the busy N19, and a long line of traffic built up behind us. The road narrowed markedly as we approached Humes. Then we were onto the sign to the campsite, a sharp right down a narrow village street and I had passed it.

'Let's go to Langres'

It seemed a good decision. We could see Langres spectacularly perched on the hill ahead, encircled by walls and towers. We wound up the hill watching eagerly for signs – initially *'Centre Ville'* then I spotted *'Camping'*. We took the left-hand lane at the top of the hill, went round the roundabout, turned right and were faced with a dark, dank looking tunnel through the town's ancient stone walls.

'Hell, can we fit through it's arched do we have enough roof clearance?'

No width or height was indicated. I drove through, both of us ducking our heads in a reflex motion but we emerged unscathed and into the town centre. We took a sharp left in the pretty gardened square and a friendly *gendarme* in his van made way for us (had he seen Boris on our A-frame, was he thinking about its legality?), then we turned right round a corner and realised we had just missed the entrance to *'Camping Municipal'* on the left and of course couldn't reverse. We had to go round again.

Another tunnel appeared; was it narrower than the last one? We were through and leaving town. A roundabout hove into view, we went round it and headed back to Langres. We went through the first tunnel, hoping we hadn't been lucky the first time and back into the square where I saw the old men sitting on a bench outside the local bar raise their walking sticks as we passed them again.

'Here's another one of those things that went through before.'

'Sacré bleu!'

This time we went straight into *'Camping Municipal Navarre'* – the site looked like a spacious woodland glade but was perched on the town's ancient walls, but where should we park, as it looked random? I took a walk round the site but decided to stay where we were. We spotted an electric post with two cables attached and a third three-pin socket free. Our electrical tester showed the polarity was reversed but our reverse polarity plug was two-pin and wouldn't fit. I decided to do without an electric hook-up. We settled in and started the gas water heater for the first time. Eventually water from it dripped down the outside of Katie onto the rear wheels and tyres so we turned it off. (*Later I read in the manual it was supposed to do this.*) It was time to take a break after 210 miles plus a tour of the town. The camp office opened around 5 p.m. and I queued to pay the 7.15 euros charge, remarkable value probably because it was a municipal site. From then on the site filled considerably.

While Gail took a rest from the journey I took a walk into Langres – and checked the leaving route to see if we would have to go under another tunnel. No, *'Toutes Directions'* took you outside the walls without going through a tunnel.

Langres was everything I had looked forward to from a small French town: open squares, narrow streets, bars and restaurants and the sun was shining. I found the walk quite uplifting. The eighteenth-century philosopher Diderot was born and lived here for sixteen years and it was said would not see much change if he returned. I walked around the ramparts and then back to the campsite virtually on the ramparts.

The next day with no overnight stop booked we decided to head towards Lyon and see how far we would get with possible stopping places at Mâcon, Crêches-sur-Saône and Lyon (Dardilly) itself.

Katie's window and front were cleaned before departure. Then we panicked – how could Gail dry her hair without an electric hook-up? I didn't think the 1000-watt inverter could cope with her hair dryer – the generator started effortlessly and we had solved that one with abundant power flowing to all sockets.

We found a way out of Langres without going through another tunnel by following signs to *'Dijon'* but at the top of the road we came across hold-ups due to road works and we needed to negotiate some narrow diversions. Eventually we were clear and back onto the A31 and making good progress.

We completely skirted Dijon, passing by evocative Burgundian wine towns such as Gevrey-Chambertin, Clos de Vougeot, and Nuits-St-Georges before joining the A6 just before Beaune.

'Autoroute du Soleil' – said the blue signs, this is what we had come for and soon we were going past Mâcon and realised we could get much farther down the Rhône Valley and beyond Lyon. We probably became a little too comfortable at this stage and did not anticipate driving through Lyon, which had much more traffic and poorer

quality roads than the exemplary ones we had been on. But here I pay tribute to the French driver. Impossible, you say. The lane discipline shown by French drivers was far superior to that in the UK where on a three-lane motorway centre lane, straddlers' pride and ego prevents them from moving over into, God forbid, the slow lane. The French will always pull over and are so keen they may pass within finger-touching distance of your front headlight.

After stopping at an *aire de service* we decided to go for a campsite south of Valence, near Crest off the D93 at La Roche-sur-Grane in the Drôme Valley. We exited at Junction 16 and crossed the N7 onto the D93, a typical French road with poplar trees each side. Making a right turn at Grane, we followed the new but narrow road for trucks into the village square where we spotted a signpost, turned left then right at the next signpost, and found we were on a narrow country road (D113) about Katie's width and a fear of approaching traffic developed. We leant forward desperately trying to find the turn to the campsite, as we would have no chance of turning round on such a narrow road.

We spotted the site on the left. Entry was achieved by turning left down onto a parallel lay-by-like road to the gate with a large sign '*Camping Les Quatre Seasons*' above. I stopped well short of the gate. As the road inside the camp also ran back parallel to the lay-by road there wasn't any way we could make a U-turn in with or without Boris on the back. Gail checked with the office and I did a walkabout to see whether we should even go in.

There were a few pitches on the top road (each of which had a sheer drop to those below) amongst which was pitch seven that had a motorhome service point opposite and better access. I returned and indicated to Gail we should unhitch. The only way through the gate was to drive past, reverse in, and drive down the road in reverse. We did this; reversed past pitch seven then went in at an angle, inching forward to get Katie's rear off the road but not as far that we fell over the front edge.

Gail fetched Boris and parked him on the next pitch. I chocked the wheels with our jack support boards. We couldn't get a true level and I was reluctant to jack Katie's rear up too far fearing she might lose grip on the grass and disappear over the hill. The slide was slid out, electrical hook up went smoothly, and a cup of tea called for. The kettle started but then stopped – we had blown a fuse – we tried another outlet on the other side of the pitch with the tester, it was live but also blew when we hooked it up to Katie. *Monsieur le propriétaire* came but couldn't mend it and we said we would manage without electricity and received a refund bringing the cost down to 13.40 euros for the night. We would rely on the inverter, which converts our 12-volt battery power to 240 volt AC.

At last we had a cup of tea (from the gas kettle). The TV dish whirred and whirred round and round, up and down, adjusting and adjusting until that lovely click, click, click noise started and I knew we were locked onto those satellites deep in space. Another day was over and 250 miles achieved.

A walk around the site revealed it as being pleasant, particularly if you were a suicidal stunt driver and could drive your vehicle round the hairpin bend beyond us and down the 1 in 8 hill to the lower terraces. The facilities that included a swimming pool and small cafe were excellent and clean but empty – no one else had selected '*Camping Les Quatre Seasons*' for their weekend revelry. The next day was Saturday but how were we to discern this? Every day was the same.

There was no doubting the purpose of the motorhome service point on the banking across the road immediately behind us as it had a toilet seat over the hole. Before departing the next day we decided to use it to empty Katie's waste tanks. I reversed Katie out carefully but couldn't get her close enough to the drain for the length of waste hose we had and had to turn her round, thereby putting her dump outlet, located in her rear, last-but-one locker on the driver's side, next to the drain.

The dump station toilet seat was slightly uphill. We donned our rubber gloves. Our blue sanitary waste hose was a 15-foot snake of flexible 3-inch diameter corrugated plastic pipe. I removed the cap to the waste outlet in Katie's locker hoping there would be no surprises and the valve really was holding everything back. The hose was connected securely by a rigid bayonet fitting (otherwise it can do a good impression of a python with diarrhoea) and with Gail in a crouching position at the other end holding the hose down the drain, her head in the air as far removed from potential odours as possible, I pulled the black water release valve and our treated sewage surged down the hose and up towards the drain. Now those with a better command of physics and the laws of gravity will understand that at this point some of the effluent decided that the uphill stretch was asking too much in kinetic energy terms and took advantage of an easier exit through a myriad of previously unknown pinprick holes in the plastic waste hose to escape in an impressive fertilising fine spray onto the surrounding grass.

We had no alternative but to let the grey water discharge, thereby rinsing the pipe of any residual brown trout and gave a further watering to the surroundings. We detached the hose and walked it to the drain to empty it then rinsed it and wondered what we would do in the future. It was unlikely that campsite owners or other campers would want the benefit of our newfound muck spreader and where would we get one for an American RV? The exercise had also shown us that dumps might not be that easy to access, needing to be within 15 feet of Katie's locker, also this one had been up hill adding to the difficulty.

We filled the water tank using our blue water hose specifically for potable water. This was simply done just like filling the petrol tank but making sure it was the right one as they are similar. I reversed towards the gate. Our target was St-Rémy-de-Provence and the Pegomas campsite making for an easier day.

As we left Avignon behind, a significant moment occurred – Gail announced we were on the page in our Michelin map book that also had The Mediterranean on it – hurrah we began to feel we were making progress in our dash for the sun and good living.

We were managing *péages* pretty well, tending to follow trucks through the outside lane where they had cones on the right and one concrete block protecting the attendant's booth on the left. At the exit *péage* for St-Remy we did this again but also noticed that the lane signboard didn't have the usual green cross but a lit-up diamond. I moved in a little way then realised that no one was in the kiosk to take our money – emergency flashers on – thankfully trucks behind noticed and took another lane. Gail descended and talked to the attendant in the next kiosk and eventually someone arrived to help. We were in the lane where trucks used credit cards. The attendant put in our card, gave us a dirty look for having disturbed her lunch, and we proceeded normally.

We left the A7 beyond Avignon at junction 25 and went onto the D99 towards St-Rémy. Tall poplar trees lined the whole route of this busy typical French single carriage road with a deep ditch at the side. Apparently we had Napoleon to thank for the serried ranks of trees planted along France's roads, as they were to provide shade for his marching troops. At any other time this would be scenic but now concentration was required.

Entering the St-Rémy by-pass, we took a signed left at the second roundabout and started leaning forward anxiously to find the campsite entrance. It was on our right, but we quickly realised it was at too sharp an angle, the gateway being virtually parallel to the road itself, so a sharp U-turn would be needed and couldn't be done. I turned right and parked in the narrow suburban street running at right angles to the gateway.

We unhooked Boris, reversed Katie and drove in. No one was in the office. The site had an open dirt area then a narrow lane with high hedges off which came narrower lanes with pitches separated by higher hedges. It was dark because of the pine trees and had a run-down feel. Gail walked ahead, walkie-talkie in hand. The pitches didn't look solid and had wheel ruts in the dried earth. I crept Katie down the lane, I saw one pitch we might have been able to reverse into but couldn't as the trees were too low. We spotted two pitches opposite each other that might allow a manoeuvre.

By now a crowd from the caravan and mobile home village had assembled for the impromptu afternoon show in the way crowds gather at an accident or vultures at a kill. We hadn't realised how big Katie was compared to other European motorhomes and were unprepared for the attention. In our minds we had expected lots of other American motorhomes but hadn't seen any.

I began to get the feeling I might get Katie trapped and the crowd had also sensed this possibility which would make the show even more entertaining and something to tell their grandchildren about. I decided to withdraw and set about turning Katie round at a crossroads of narrow lanes by the concrete toilet block. The crowd was almost at fever pitch anticipating Katie jammed up against the low tiled roof and drew dangerously close but were disappointed as I safely executed a six-point turn and they drifted back to their afternoon television.

Boris was hitched back on. An Irish couple tried to persuade us that another site on the other side of town had big pitches but we would have had to drive through town and they said it was narrow.

It had been disappointing. I had viewed Camping Pegomas on the web, it had looked really good and the *Caravan Club Guide* said 'excellent well-run site'– a lesson learnt – we would stay out of a site until we had seen it for ourselves and would not expect anyone to be there from 12 noon to 2 p.m.

We were on our way again but to where? Two of the three campsites we had visited had been difficult to access; this was not something we had anticipated. Would other campsites be equally tight – were we just too big? I picked a name off the *Caravan Club* map '*Vidauban*' well into the Mediterranean area beyond Aix-en-Provence, St-Maximin-la-Ste-Baume and at a junction between the A8 that we were on and the N7. It was impossible to anticipate the situation we would find and a degree of anxiety prevailed.

After the junction we took the N7 east and proceeded cautiously to Vidauban, rush hour traffic building up behind us. I pulled over to ease the imagined congenital frustration of the French drivers and a replay of the *Duel* sequence. But still no '*Camping*' sign appeared and we were now in Vidauban and having to take the narrow one-way system through the town centre with old four-story terraced houses leaning in on each side of us. Stress levels began to build. We were then out of the other side in two-way traffic and going even slower when, before an iron bridge, we spotted the sign and made a left turn across the road getting a few derisory hoots from behind.

Following another sign at the top of the hill brought us to a narrow lane that came to a T-junction and a sign '*Impasse du Camping*'. Did it mean you could pass if you were camping or it was forbidden to campers? The road was a suburban street with bungalows on each side each with nicely manicured front gardens. I saw the possibility of getting stuck down it and being unable to reverse.

Gail did it on foot and returned to say it was OK. I drove to the camp entrance where welcoming fairy lights dangled across below our roof height. The proprietress came to see but made no attempt to help and did a good impression of a Gallic shrug. Word reached the rest of the campsite and another appreciative crowd assembled at a cautionary distance. Dogs even came to watch.

Gail climbed up the rear ladder onto Katie's roof and I drove slowly forward while Gail lying on her back slid along the roof and fed the fairy lights over Katie's roof mounted aerials, vents and air-conditioners. What a brave girl. As she hadn't fallen off the roof and we hadn't snagged any lights causing a multi-volt explosion the crowd melted away and the dogs resumed their games.

We were on a pleasant, spacious and flat grassy site – '*Camp Municipal Les Bords de L'Agrens*'. We could park where we wanted and hooked up to 6 amps, the electric kettle worked, as did the satellite, despite the many trees. Because of the diversion to St-Rémy it had been a long day and we had expected it to be much easier but here we were close to the Mediterranean, one of our dreams, and we could now relax.

Soon after thunder and lightning started and lasted throughout the night requiring me to get up to lower the satellite dish, Gail once more fearing incineration.

It was Sunday September 28, and we discovered the campsite would close on September 30. We had to find another.

We decided to go to St-Tropez in Boris, and anticipated a beautiful drive over the hills to Ste-Maxime and along the coast but it rained all the time and as we parked in St-Tropez there was a deluge of such proportions that we were trapped in the car by rising Niagara waters for 30 minutes wondering whether lifeboats rescued from flooded car parks. Many months dreaming of tropical sunshine on the Mediterranean coast, deciding whether to walk the beach, have a seaside lunch alongside the glitterati, mooch the expensive shops pretending we could buy – who would know us, we could be anyone we wanted – were dashed. We were a couple trapped in a little white car watching rubbish bins float past. It continued later as we went to Ramatuelle looking at campsites on the way.

Have you ever looked for a long cherished place hoping on the one hand that it would be exactly as you remembered it and on the other knowing that the place will now be something completely and disappointingly different, but you still look? We played guessing games with the map because we could picture a marvellous Auberge we stayed in when we last visited many, many years ago and then argued about every turn as to who could remember where it was. We didn't find it. We diverted to Le Lavandou but the campsite was too small.

On the way back we stopped off at *'Camping Les Cigales'*, a site near Le Muy. This was a possibility – a group of young people were looking after the large and modern reception – we wandered the secluded site – it had good-sized pitches set amongst pine trees and sand hills. Tree branches limited access to the lower pitches and those on the open top road shortly after the entrance looked the best. The charges were reasonable at 14 euros per night with 6-amp electric and we saw a water hose alongside the pitch.

Having only one day to find another site was a strange unsettling sensation. We wanted to enjoy the real France but after a week I would also have welcomed a chat with other English campers to share experiences or garner information but we hadn't met any.

After an evening's research we set off in Boris to see around six sites to compare with Camping Les Cigales. The weather was good. We went as far as St-Aygulf through St-Raphaël to Agay and back but none appealed or were practical.

'Stop stop!'

I had spotted a yard full of plastic pipes of every size and colour just off the N7. I went in and showed our blue waste pipe to the man who said *'non'* – but how could this be – you've got a whole yard of the bloody stuff outside? In disbelief we went into the store yard and found a huge number of rigid plastic pipes used for below soil services. He came out and waved in a told-you-so manner.

Mini-skirted prostitutes plied for business on both sides of the N7 near Vidauban – I don't know why I mention it but they were hard to miss as was the soiled looking mattress by the hedgerow –surely they didn't use this? Girls it's about presentation; a little investment here could reap rewards.

After a cup of tea back in Katie we tried one more site less than 5 miles from Vidauban without success – Les Cigales it was then. We hooked up Boris ready for the move the next day and set off for an evening walk to town but had to come back for the camera I'd left in Boris – this set off Katie's alarm. We apologised to the lady in charge of the campsite – she said it had gone off *'plusieurs fois pendant le jour'* – several times during the day – how embarrassing.

It was Tuesday September 30, and the Vidauban site closed that day. In many ways it had been perfect, with flat grassy areas, mature trees that didn't interfere with manoeuvres, good access apart from the fairy lights and value at 26.20 euros for two nights including electric. We regretted not being able to stay longer and relax. It was a hassle to have to move on. We didn't need to hurry but as we finished breakfast we were one of only three campers left on the site. Gail mopped Katie's roof so she wouldn't slip when she rode on it out of the site entrance. I swung Katie round and with Gail lying on her back as before I drove slowly and as smoothly as I could out the gate and Gail fed the fairy lights over the roof obstructions.

I descended from the cab and offered, *'Je m'excuse'* to Madam for the *'bruit de l'alarme.'* She kindly said, *'Pas de problème et bonne journée.'*

We were away, through a couple of tight turns and onto the RN7 turning an immediate sharp left under the 4.1-metre railway bridge, our aerials rattling hard against it (Katie's roof is 3.65 metres high).

A short drive along RN7, a cheery wave from the prostitutes on early turn, and we had soon driven onto the

N555 headed towards the motorway péage but ready to take the sharp turn right before it into Camping Les Cigales. It required a swing out into the left-hand lane of the dual carriageway and I simply knew an impatient French driver would squeeze through on my blind side; couldn't he see the fairground of flashing amber illuminations on Katie and Boris indicating my intentions or did he think: *'zees rosbifs zey don't know their left from their right?'* We drove down the road to the barrier of *'Camping Les Cigales'* all of 9.3 miles from Vidauban.

The cavernous office had a desk in one corner, a small table in the middle with brochures, an Internet facility at one end and acres of unused space and no campers. The young girl and guy were busy with important computer things and we apologised for interrupting. It had that end of season feel. We signed in then enthusiastically headed for pitch 190, the one we had previously selected. This was a recently developed pitch and the JCB has simply dug away the sandy soil, flattened the pitch and dumped the spoil on each side to create banking just as a grave before they lower the coffin in, only shallower. We unhitched Boris and drove Katie onto the pitch remarking on the great view through the windscreen of the surrounding fields, pine forest, the more distant hills and if you looked closely enough the motorway *péage*.

We ran Katie's rear dual wheels onto the mats we carried to prevent her sinking then lowered her four hydraulic jacks onto wooden blocks to level her up – but we couldn't do it without almost lifting the front wheels off the ground – something I didn't want to do in view of the soft sand we were on.

We tried again, creeping forward onto soil that appeared to be a bit higher, but possibly softer, and ran the front wheels onto mats and managed to level. With overnight rain forecast we hoped we wouldn't sink.

The electric supply had a modern three-pin socket but our tester registered *'No Earth'*. Examining the box showed an incoming earth, and earth wires to each socket but nothing connecting the two. I tried the older electric posts farther down the site in the woods and they tested fine but were a quarter of a mile away from us. I went back to the office.

'It has worked for everybody else.'

Gail, convinced I hadn't communicated effectively, went again later and came back to say someone would be coming – no one did, not that day or any other.

We also noted that the water pipe we had previously thought would give us a local supply, was just a hose lying on the ground – not a tap in sight.

We decided to use our own resources for electricity. I switched on the inverter to power up the TVs; but moving forward had put us between two trees one blocking communication with Astra 2 out in deep space. A walk round the site showed how extensive it was with a beautiful swimming pool as well as an 'infinity pool' disappearing over the edge into the forest with its roads, hills, paths and hidden shady glades. A few residual Dutch caravan campers nestled almost hidden in the woods. We liked the site and could relax there in the Mediterranean sunshine until the end of October. A large block building provided the toilets, showers and food preparation areas and an enclosure at the back a giant toilet for waste disposal. I silently worried, but said nothing, about how I was going to get Katie anywhere near it to discharge her tanks with a waste pipe we didn't have.

Le Muy

I was keen to re-discover the atmosphere of Provence we had enjoyed on two freewheeling trips by car in the 1970s and 80s when we made unscheduled stays in *Chambres d'Hôtes* often gleaned from Arthur Eperon's, *Travellers France*. Both visits had left a lasting romantic impression, but that was all, I could only remember one or two actual events or places. Gail seemed able to recall much more, was senility taking hold?

We started immediately with an exploratory trip in Boris west to the village of Les Arcs Sur Argens that proclaimed itself an attractive village dominated by the heritage-protected eleventh-century medieval citadel and as the wine centre for the Côte de Provence. We parked in town and eagerly set off on foot through the arched clock tower entrance in the fortified walls and up the narrow winding cobbled streets. Save for the odd cat asleep on a sun-baked stone doorstep we found a near deserted picture-perfect medieval village in honey-coloured stone with much restored houses, arched doorways, flower-bedecked windows, vaulted passages and old wooden doors with sculpted panels and brass hands for knockers. At the top of the hill, we found a marvellous hotel building (presumably the Citadel of old) and Saracen tower with magnificent views over the surrounding town's patchwork of terracotta-tiled roofs, vineyards, and in the distance the Val d'Argens and Massif des Maures Mountains. The cottages clustered tightly round the hotel have all been converted to hotel bedrooms accessed by the front doors from the cobbled square but the near perfection of the conversions made it feel sterile except that it wasn't because of all the doggy doo piles. A smartly dressed couple, he in his white casual trousers, dark blue, short-sleeved shirt and white peaked hat, she in a flowery summer dress and sun hat walked their two dark brown dachshunds. Then he stepped into a poo pile with his pristine white gold flecked trainers. I struggled to suppress my wholly inappropriate mirth.

Back in the main square at the base of the old town we had two *café crèmes* and a *croque monsieur* French style, sitting no more than two feet from speeding traffic. Another couple, he in his pink shirt, Harris Tweed jacket and beige slacks, she in her sun dress and blazer were British caricatures and to confirm it he started reading the *Times*. Then the wife of the man with the dog poo on his shoes emerged sheepishly from the toilets having been to clean it off.

Although we aspired to shopping at markets for individual fresh items grown lovingly in a small farmer's lot, harvested by his children, and beautifully presented by his adoring wife we were short on food, so we stopped at *'Ed's'* supermarket in Le Muy. All supermarkets place the fruits, vegetables and salads as you enter, a strategy designed I understand, to give you that market-fresh feeling. *'Ed's'* could have been in Iceland (the country, not the shop) as it was freezing inside and I soon lost the will to be interested and wandered around as Gail did the shopping.

I noted that people were weighing their produce on scales where a label popped out of the side that they then stuck on the sealed plastic bag, a system alien to our usual Isle of Dogs Asda. I hastily informed Gail, what confusion would have reigned if we had turned up at the till without our weighed and priced produce?

We looked at the meat counter; the butchers with bloodied aprons stood behind sharpening their knives with that special relish to do our bidding. Despite our quasi-medical backgrounds many parts of the animal anatomy were unknown to us so we made a few guesses and plumped for a smoked chicken. The French wines looked like bargains so we bought some Californian wine that looked ridiculously cheap at 1.20 Euros a bottle. As I do not drink alcohol, Gail would see how it was.

Cheeses from solid yellows to oozing creams and blues were ready to be sampled, sliced or scooped. The cold was affecting us; I had lost the will to live, and wanted to get out. At the till we maniacally loaded up the conveyor in an effort to speed up the shopper in front who lingered to tell the cashier of her latest *'maladie'*.

'Pas du lait ou de l'eau! Dedans le chariot!'

The milk and water were too heavy; they must go back inside the trolley. What feeble French conveyors are you using? You may have fancy printing weighing machines but our Asda conveyor will tolerate a good load of semi-skimmed and Highland Spring. Maybe if you increased the ambient temperature the conveyor wouldn't creak so much?

We went back to Katie and had one more try on the satellite – no it would be CDs again that evening. We enjoyed our smoked chicken in a salad. For our first night at Le Muy it rained, drumming loudly on Katie's aluminium roof half waking us – was Katie's front sinking – was that the reason I felt I was sliding forward – or was it my imagination? Back to sleep.

By nature I'm a worrier seeing the glass half empty but this steels me to drive onwards relishing the challenge I've created inwardly for myself whilst presenting a calm exterior a sort of self-preservation that I reason comes from being an only child and talking to myself a lot. With Gail's reservations about the purchase of Katie it was important things went well. With a distinct sense of unreality I found it hard to relax and a variety of thoughts and questions churned in my mind: had we really done that journey, would we ever find a perfect campsite and get settled, what was wrong with where we were, did we want to settle, what might go wrong with Katie's complex machinery, how could we dispose of our waste without a working sanitary hose, when should we go back, was it right to stay away from family at Christmas, would we make it back, wouldn't all the campsites have closed – they all seemed to be closing even in the South of France?

We succumbed to our desire for the real French market experience with little stalls stacked with a rainbow of fresh vegetables and fruit, strange meats, fresh baked bread you ate as you walked around, oozing pungent cheeses from sheep and goat's milk, pâtés, foie gras, confitures, olives, figs, wine and stallholders in berets and striped blue and white T-shirts but perhaps not dropping Gauloise ash over the produce. Draguignan market beckoned and close to the centre of town and the main road we found a large park or square (the *foire)* crammed with the stalls and people of a busy market that disappointingly sold clothes, rugs and bedding. Unsure what to do, we wandered through the pedestrian zone into the centre of the old town where in another square (Place de Marché) was a typical French market scene of tightly packed stalls with colourful striped awnings offering wonderfully fresh baguettes, olive and nut bread, garlic, olives, olive oil, charcuterie, fish and cheeses with shoppers, their dogs and children pulling excitedly at their arms, filling their wicker baskets whilst swapping friendly banter with stallholders.

We were late and much of the produce had been sold already but the atmosphere we sought was there and as we drank our coffees at the outdoor café in the sunlit market square I felt we had arrived and started to relax and enjoy the real France. I turned to Gail; yes she had also noticed the two other tables were occupied by Brits.

The nearby small village of La Motte provided an exciting drive and after two circuits of the one-way system designed to give the local bus driver half a chance of not destroying the overhanging medieval houses as he turned the tight corners, we managed to park. As we had done our lap of honour I spotted the all-important WC sign. We found the '*cabine'* in the *boules* square. A sheepish young couple came out together. *'Quelle surprise!'* The Ladies was jammed or locked so it was all for one and one for all. I stood guard as Gail used the only croucher. (I never adapted to these things

– when needs must and it was pants-down time I hated every minute, adopt the position, watch your trousers don't touch the floor that was always wet, make sure your bum is out far enough so everything goes down the hole and not into your trousers, why can't I be quicker, talk about unfit, thigh muscles ache, oh hell I can't reach the paper I brought and put in my trouser pocket. How do eighty-year old ladies manage?) We washed our hands at the convenient fountain in the square then went up the hill to the clock tower – Gail posed with the dummy (no a real one) pulling the flower cart.

Stone arches spanned many of the narrow streets. Two young girls were having their lunch on top of one of the arches – this was their patio – flowers spilled down the arch and from window boxes everywhere – heavenly. We took a photo. Would it convey the real beauty?

On the way back to the car we were struck by the size of the school provided for this small and ancient village – a large modern building surrounded by sports fields, tennis courts, a playground and car parking and subsequently found this focus on education a common occurrence in France.

We were distracted by heady vinous smells that emanated from the 'Caves' of the wine cooperative that had a glossy double-glazed smoked glass entrance at the front but behind looked more like a factory. Down the lane and to the side of the buildings a yellow JCB was bucketing up red grape and vine residue from a mountain stacked against the walls of the 'Caves' into an articulated lorry. Once full, the truck pulled out into the lane to be immediately replaced by another eager to get its load.

As we drove back on the N7, we spotted a place selling European camping-cars and accessories. More in hope than expectation we explained to mademoiselle that we wanted a pipe for the toilet waste. She soon had exactly what was needed, a short length of grey flexible hose with adaptors at each end so we could couple the waste outlet directly to the waste water roll tank (the toilet dolly as we call it or honey wagon as others do) and a longer hose with no adaptors for free discharge. Once back at Katie we removed the well glued on American adaptor from the leaking blue hose and with the aid of some washing liquid inserted it into the new longer hose and clamped it on. When had a sanitary hose created so much pleasure? We were ecstatic; it was one less thing to worry about.

We sat out in the sun, enjoyed a cup of tea and vowed that tomorrow we wouldn't do anything at all but relax. In the evening we listened to French Radio – would we soon be native speakers?

As we were relying on our batteries for power (there are three 12-volt deep cycle batteries available for this and separate to the engine battery) I kept a careful watch on the charging by the roof-mounted solar panel. An LCD display just inside the entrance door provided information on battery status and the charge being produced. My concern meant I developed the habit of looking every time I entered or left Katie. So far that day there had been no charge but as the sun rose it ticked up to 0.8-amps.

We set up the chairs, the table, and a lounger and sat in the sun or shade as our wont took us. The sun was sufficiently intense for sun cream and I sat outside listening to French language tapes that I found had a remarkable sleep-inducing effect. Gail decided to clean Katie inside as I dozed.

'ANTS!' she yelled, jolting me from a deep dream.

They were crawling up Katie and in the driver's door. I found the ant spray we had prudently bought and sprayed the wheels, the wooden blocks under the jacks and anywhere else the ants might decide to scale. Investigation of the surrounding soil bank showed some huge ant runs and nests so they received a real dousing.

The rest of the day was blazing hot. Then the sun, a golden orb, became a fiery red and sank slowly over the forest. The solar had worked well all day with 5-amps charging the batteries fully; what a relaxing time we are having.

Later that evening we were sitting in Katie when we heard a puzzling clicking noise. It was seven days since we had discharged the waste tanks and filled with water at Grane, so we decided it meant one of the waste tanks was full and the bathroom computer panel confirmed it was the grey water tank. We envisaged our morning shower water backing up through the drain and flowing out into Katie. *(On return to the UK Nigel told me there is no such alarm.)*

It was pitch black outside. I switched the locker lights on from inside Katie, and removed the 40 litres blue plastic toilet dolly (more on this later) from a locker and positioned it on the ground under the waste outlet. I wriggled one end of the new short hose onto the dolly and the other up through the hole in the waste locker and onto Katie's waste outlet then opened the grey water valve to fill the dolly. We re-checked the tank levels on the computer panel. The grey was now two-thirds full but I also noticed we were down to one third on fresh water. We discussed what we would do about fresh water and waste. If we were to stay we must have a strategy, seven days was about our maximum before attention was needed.

Next morning I was out early to empty the grey water from the dolly, the first time I had done this – like barbecuing, it was a man's job and I enthusiastically wanted to show Gail it wasn't really a problem simply a job to do. It was a long way to pull it along the campsite road then half way down the hill through the trees to the toilet block. With those little black plastic wheels on the grit of the road it was a noisy beast so the few campers remaining knew exactly what I was up to, but twitched their curtains back to check.

The 'WC Chimique' (chemical toilet) was located up some steps to a path at the side of the toilet/washroom block and at the rear, in a cave-like tiled enclosure. It was like a big toilet with a five-bar metal grid hinged from the back that could be lowered down on to the top of the bowl to provide support. Those with smaller cassette, or 'porta potti' toilets, in caravans and European motor homes simply upended the contents of their one-day cassette into the toilet through the spout and watched in a satisfied manner as the brown trout were flushed away. Not so with ours. I heaved it onto the top of the porcelain in a vertical position and made sure the wheels were positioned for maximum stability. This was difficult when they were about the same width as the bowl, and porcelain and plastic do not provide any grip. My left forearm and back tensioned to take the strain of the 40 kilograms of the slopping soapy shower water. I was in an awkward position and decidedly out of control.

With my left hand I lowered the top of the dolly down, away from the toilet, to the horizontal so as to be able to open the cap on the lower exit hole with my right hand. Then, straining to stop the wheels slipping off the porcelain, I slowly raised the whole thing up so the liquid would flow.

It came out in a forceful, bumpy, unsteady, uncontrolled, glug-glug way, the biggest glugs reaching the back of the toilet. Previously unchallenged muscles strained to control things and I was relieved to be only dealing with shower water.

Back at Katie, I filled the toilet dolly with black (toilet) waste sludge and trundled back down the road, apprehensive as to what might happen. With the dolly vertical on the porcelain again it occurred to me that opening the air lock would stop the previous bumpy flow. As I released the outlet cap the smoothest of uninhibited flows ensued, a gusher of effluent spouted into, across and over the back of the toilet and down the slope towards my shoes.

The large unscrewed outlet cap, dangling like a camera lens cover, caught in the stream of effluent and became a wild pendulum swinging backwards and forwards in and out of the flow to create an arc of spray just like a lawn sprinkler. Panicked and to stem the flow, I tipped the dolly back but it started to slip off the toilet so I could only get it back a little. I closed the air lock. The resultant glug-glug flow in combination with an outlet cap in full pendulum mode created a new and wider spray pattern. I grabbed the effluent-soaked cap to stop the spray and desperately tried to keep the dolly on the toilet, otherwise all would have been lost. The last glugs finally drained out and I hosed everywhere down. Who invented this bloody system, I cursed under my breath, this can't be how it was supposed to work?

I took a coffee break. I didn't feel at all manly.

'This is impossible,' I moaned to Gail, 'to empty the one hundred and eighty litres grey and black tanks is going to take me at least nine dolly runs every week.'

'Would you like me to help?'

The moaning had elicited the sympathetic response sought and Gail and I did the final runs together, although I insisted on doing the wheeling so that curtain twitching campers retained a macho view of me (although with the quantities we were taking compared to their small overnight emptyings they must have thought we were shitting for England). We decided to retain the short waste hose on the dolly so that one of us could direct the effluent flow into the toilet whilst the other held the dolly steady and this worked better.

Emptying the tanks was one thing but we also needed water. I had seen a water point amongst the trees by the swimming pool.

As we drove up the road we must have been quite a sight because the few campers left on site all emerged from their caravans to watch. I couldn't make the left-hand turn without risking scraping a low wall so I turned and reversed down. A camping Frenchman who had just seen this large vehicle approach his pitch leapt to assist. He gave me exaggerated steering wheel signs from the front, Gail talked to me on the walkie-talkie from the rear and I was watching the mirrors and rear-view camera – too much information. I crept backwards down the hill to the water point. The Frenchman we think explained that we didn't need to have done this, as there was a water point by the entrance. We filled the tank through the gravity fill. As we went back up the hill our friendly Frenchman was waiting – he had been working out our return route so I followed his guide and we turned Katie round and returned to our pitch.

When we got back we saw that the water tank was only two-thirds full and reasoned it was because it had been filled on a slope. All was not lost however as we were able to angle Katie across the pitch, the satellite missed the tree, and *Voila*! BBC. What's more, Katie was nearly level.

The next day the fresh water was down to one-third full. The thought of manoeuvring to fill up again and still ending up with about two-thirds of a tank as well as all the toilet dollying was unsettling. We considered our options of going to another site, finding an *aire de service*, or continuing to empty the tanks where we were.

As I checked things over, a Dutchman passed by and commented on how wonderful Katie was. I explained the situation with the water fill and the electric. We discussed the electric, both of us peering into the junction box. He said forget having an earth – nothing is earthed in France; the fuses in the box are our protection along with our four tyres. I pointed out we are actually standing on four metal jacks but he said they were on wooden blocks. I pushed the plug in and we had power earthed or not. On the way out for shopping we stopped off to tell the girl in the office that we were now electricity consumers (they must have thought we were mad).

We took a Sunday drive to the coastal town of Fréjus; it was sunny but also windy. We found free parking but I was nervous as there was a man sleeping in his car that was obviously home for him and another dishevelled character sat on the kerb by the entrance.

We walked into the old town where the church was discharging its congregation into the square. The old medieval town had a good feel to it with many Roman remains, baths, amphitheatres and aqueducts and the 'Cité Épiscopale' one of France's oldest ecclesiastical buildings.

Our guidebook informed us Julius Caesar had founded it in 49 BC when it was known as Forum Julii and was then a port where in 31 BC the warships with which Augustus defeated Antony and Cleopatra at the battle at Actium were built. The port eventually silted up but was resurrected by the construction of Port Fréjus. Because of the resident vagrants I was happy to get back to the car park rather than have a coffee.

Two miles farther along the coast at Fréjus Plage, multicoloured market stalls, many manned by Africans selling all manner of watches and sunglasses, stretched along the whole length of the Plage Promenade but stallholders were loading their vans, keen to close and get out of the wind before their flapping awnings took off and became hang gliders. On the main road behind the Plage were many restaurants with a few brave souls sitting outside. We wandered along thinking the *Moules Marinière* and *Provençal* looked inviting.

Down by the marina, near the police and fire station, we found that the Aire Communale de Fréjus was not a public gathering ground but a blue roadside unit dispensing water, receiving waste drainage and supplying electric for motorhomes all for 2 euros. Could this be the answer to our exploits with the toilet dolly?

A few days later as Katie's water tank emptied and her waste tanks filled once more, I spoke to the girl in Reception.

'Is there a water point close to the camping-car?'
'Non!'
'Is there one near the office as a Frenchman said there was?'
'Non!'

99

I explained that the water tap *'dans l'intérieur'* of the camp was *'difficile'* because of *'les arbres.'* I caught a word she was saying that I recognised. 'La Motte.' She suggested the only way to deal with things was to go to the village of La Motte.

From a guide to Var we had purchased, we learnt that at 23 Boulevard André Bouis in the village of La Motte there was an *Aire de Repos Camping-Cars*.

We prepared Katie for a move and drove along the dual carriageway up the D555 then turned right and along a narrow country road across the hillside vineyards and down a steep and narrow hill into the village of La Motte.

At the bottom on the left by a little white flower-bedecked tourist cottage was an entrance to the Council Garage – where a Union Meeting was in progress. A dozen men in traditional blue overalls were in fervent discussion – probably how they could get more EU money to preserve their traditional way of life driving street cleaning vehicles. The sight of Katie did not deter them. When the men realised they had encroached on their two-hour lunch break, the meeting broke up and the men dispersed. Once they had moved off we could see five bays specifically marked for camping-cars, a dump drain and fresh water. Stopping was limited to 24 hours.

A Frenchman emerged from the only motorhome and tried to tell me something about the water I didn't understand. His wife came to the door in her nightdress. They thought the water smelled of chlorine.

The open drain made emptying through our new grey hose easy and with water available the black tank was thoroughly rinsed –it felt therapeutic. We were now in a position to fill Katie with fresh water. It was straightforward enough Gail holding the hose in the filler whilst I operated the tap. The monitor however only showed two-thirds full so I surmised that once again Katie wasn't level and told Gail I was going to drop the jacks and level Katie up if she could just hang on a moment with the hose in the filler. It worked a treat. As Katie levelled I heard a noise, best described as a cow farting, issue from her water tank followed by a tsunami wave of water that soaked Gail from head to foot.

I was delighted to find that the tank filled and the level read *'FULL'* whereas Gail's interest seemed somewhat dampened. (What we didn't realise until much later was that Katie has two filling mechanisms and if we had connected the hose directly to the *'City Water'* inlet no burping would have been required).

We had soon returned to the campsite and set up again. Properly level this time, satellite working, electricity hooked up and tanks empty and full where they should be. For the first time since Grane on September 27, we were complete and enjoyed a cup of coffee, reflected on life and realised what a godsend La Motte was going to be and would enable us to enjoy our time. Thereafter we visited every Sunday for *'eau vidange and eau user'* (water drains and water to use) but that little computer panel on the wall above the bathroom sink continued to obsesses us.

We met a number of people using the *aire* overnight at La Motte including a British couple in a Hymer motorhome who had been to Greece and were on the way to Spain and recommended Almeria. That started me thinking about moving on and getting to Spain. It was also strange to be talking to people because the campsite was virtually empty and we had few conversations.

We were well settled and attended to the daily and weekly chores. Gail ventured to the campsite laundry, I used the convenient Internet *'Borne'* in Reception for e-mail and discovered that our Internet bank had introduced a six-figure Internet security code thereby denying us access. I made expensive mobile phone calls to the service desk in some far-flung region of the world with charges clocking up. Illogically for an Internet service they had posted the PIN to our home. We agreed that a new PIN could be sent to my mother. Thinking about it there was surprising little security to get this done.

I started to clean Katie as this had not been done since we left home and was really dirty. The polish I used is called *'Dri-Wash'* an American product that as its name suggested meant you didn't have to wash first. This was handy, as many campgrounds banned the washing of vehicles on pitches. You simply sprayed the polish on rubbed gently with a *'Terry'* cloth then polished off with a clean *'Terry'* cloth. Fears about it scratching the paintwork appeared to be ill-founded.

The sun blazed down and we enjoyed many a lunch outside and lazy afternoons under our huge green shady main awning. In almost splendid isolation we spent time sunbathing at the wonderful campground pools with cooling waters that looked as if they disappeared over the edge into the pine forest and took photos to impress those at home beginning to endure the start of winter. We filched pomegranates from the trees surrounding the pool.

Reality broke through and required us to post a letter with a cheque for our accountant's annual fee. After an early lunch we set off for St-Raphaël, a large seaside town, nestled up beside Fréjus, and located halfway between Cannes and St-Tropez.

Despite being out of the holiday season, parking was difficult and after a circle round town we parked east of St-Raphaël next to the beach where all the windsurfers were battling strong winds.

In town we tried to find the Post Office but became lost and ended up at the station where we had first started and stared once more at the map on its wall. The railway, that courtesy of the TGV could whisk you to Paris or bring an escape from city life to The Riviera beaches, divided St-Raphaël. Once we had discovered (on of a map from the tourist office) the tunnel to get to the other side and thence the bus station we were comforted by an old signpost to the 'PTT' (Post Office). Whilst Gail fought with her map in the wind like some demented dervish, I reconnoitred and found another sign with 'TT' on it. Assuming some mischievous imp had obliterated the 'P' we followed this and after 30 minutes were queuing for a single 50 cents stamp to post the cheque.

We wandered through town into the old quarter looking at the church, the old town walls and supposed fine old eighteenth and nineteenth-century villas but it was not as picturesque as Fréjus and neglected. Even though a coffee out of the wind overlooking the harbour was welcome, the 5.80 euros price was expensive, a prime position but not prime conditions – a bit mucky.

We walked back along the seashore struggling against the wind examining restaurant menus as we went and agreed that Fréjus had won the beauty parade.

That evening we enjoyed a wonderful hot dinner but had no TV as the wind was too strong and a risk for the satellite dish. Katie was rocking and with the trees bending Gail was worried. We hoped for better weather the following day and planned to visit the Gorges du Verdon.

After experiencing a windy night, we woke with the sun breaking through and calmer conditions. We would go on our tour to France's equivalent of The Grand Canyon that was said to provide drivers with a white-knuckle challenge. We packed fleeces as we were going to the mountains and didn't know what to expect. Gail drove us first to Draguignan up the N555. Draguignan was busy and we negotiated our way through to the D955 and had already started to get some S-bends and then hairpins. We stayed on the D955 after it had divided from the D21 and D71 and become a minor road. Our way was signed *'Gorges à Droit'* and took us to the north side rather than the D71 going to the south rim. The scenery was magnificent with deserted heath and hills and at Trigance we spotted a magnificent chateau on top of the rock face.

Soon enough we joined and turned west along the D952 from Castellane and were heading into the Gorges which run for 21 kilometres with the River Verdon at their feet, sometimes slow flowing other times rushing. We stopped in the warm sunshine to take the obligatory photos at '*Point Sublime*', where the river plunges into the narrow rock walls, and doesn't escape until it comes out the western end before flowing into Lac de Ste-Croix.

The road onto La Palud-sur-Verdon was exciting with rocky outcrops sheer drops and a lot of rockfalls resulting from the overnight winds. We enjoyed it immensely. We stopped for coffee in La Palud-sur-Verdon, a small village with narrow streets that was once a potters' village with a tradition for local products. Here, our guidebook told us we would find trout and crayfish, baby lamb with *Provençal* herbs, lavender honey, truffles, ham, and country bread. Sitting outside in the clear mountain air and sunshine we extended our stay to sample lunch and requested the local *Plat du Jour* that turned out to be chicken and potatoes.

We unintentionally took a tour of the town three times as we hunted for the *'Route des Crêtes'*. It turned out to be the one we were originally parked on. The road was spectacular, with lots of photo opportunities of the 800-metre drop, but it lasted only 8 kilometres and was then closed so we had to turn around. The weather was perfect with clear blue skies.

We headed for camp satisfied with a great day and sat out in the sun to enjoy the last rays of the melting sun.

'What a wonderful camping-car.'

Two Dutchmen had stopped to admire her whilst I was doing some polishing and I answered their questions about Katie and how she worked. They were also retired but unlike us had to go home or the grandchildren complained. They asked if we were going to Spain, as it was something they would like to do.

We set off for a tour aimed at Lorgues to the northwest of us. We went east on the N7 then up the D10. Lorgues was busy and we parked in an off street park alongside the high street. Gail worried about Boris not being parked accurately on the parking lines – you could hardly see them and this was France on a Saturday.

We liked Lorgues instantly and read later that it was something of a gourmet stop, truffles being a speciality. There was lots of activity, people sitting out at many restaurants and cafes. We bought postcards with maps of the Var region, thinking that parents would be able to see where we were, as well as some follow-up cards of the Gorges du

Verdon. We wandered round the narrow winding old streets, looked at restaurant menus and promised ourselves a return visit as it had a good feel to it.

Rather than eating gourmet-style, we had brought some food with us and we sat in Boris and ate salami speculating what it might be made of and pig's face was mentioned.

We extended the tour by going farther east on the D562 to Carcès. The road ran alongside the River Argens and provided a typical *Provençal* scene of farmland, vineyards and woods.

In Carcès I directed Gail onto a little byroad to the Lac du Carcès. It looked attractive on the map in the way that blue coloured lakes and green coloured forests are. The 'road' was a narrow, single, rock-strewn track with low trees, one of those where you begin to wonder just where you will end up and wish you'd never started. The lake only came into view at the end where the road stopped. It was a man-made reservoir and without character. To assuage my guilt at bringing us down the track, I drove us back to the main road where Gail took over again.

The lake was no better from the other side and further on many mines blighted the area. I decided to head for the Abbaye du Thoronet in the hope that this might be of interest and it turned out to be well worth the visit and they provided a guide in English. The abbey, built between 1160 and 1175, was home to Cistercian monks and lay brothers who produced olive oil and wine but in 1660 the place fell into decline. In 1790 only seven monks lived there and the abbey buildings were in near ruination. Restoration started in 1873 by the architect Revoil and recommenced in 1907 by his successor Formige. The restoration was incomplete but it was perfectly possible to envisage the life of the community.

We woke at 07:30 on Sunday October 12, and Formula One was on the TV in the bedroom, would Schumacher get his sixth world championship? – He did.

We decided that we would fulfil a promise made last week to go to the market at Le Muy, reputedly one of the best in eastern Var. It became busy as we drove into town then as we arrived in the centre it was jammed and traffic was diverted to the outlying car parks. We walked back into town. The stalls we first encountered were selling clothes but as we progressed to the top end we came across stalls selling the vegetables, garlic, oils, breads, meats, rotisseried chicken and some with giant pans full of simmering couscous and paella.

The atmosphere was exactly what you would expect of a *Provençal* market and the sun was shining nicely down on us. We had coffee virtually on the main street being lucky to get one of the few available tables. We bought *The Sunday Times* that had been printed in Marseille and wondered whether it would devalue our trip, wasn't British TV enough – shouldn't we have been doing French crosswords?

The market was quietening but vivid colour prints of *Provençal* scenes in horrific blue-white frames at 20 euros apiece were walking off the stall. Almost all of the stallholders seemed to have cleared their goods and even the couscous pans were being cleaned up. We bought warm bread with olives and anchovies for Gail and pizza with tuna for me.

When we returned, the warmed bread and pizza full of herbs and olive oil provided a delicious lunch. After our usual Sunday trip to discharge and replenish tanks we luxuriated in reading the Sunday paper.

People were leaving the campground at the rate of two or three a day and few were left. Our tourist enthusiasm however was unabated and Tuesday was market day in Lorgues. We luckily dropped into a parking place on one of the approach roads. As we walked up the hill (avoiding the dog poo) Gail was engaged by a stallholder selling tapenade and after taking the taste test bought some made with black olives. I commented that it might as well have been caviar at the price– the stallholder thought that was funny – I didn't.

The market was great but we only bought bread with *noix* (walnuts), and then wandered the rest of the stalls enjoying the sights and smells of the local produce. As lunch approached we decided to return to the campsite where we took pleasure in the bread, cheese and tapenade with a glass of wine for Gail. The afternoon was spent sunbathing; what a life we had, and the next day was my birthday.

Katie has two wheels at the front and four dual wheels at the back. The tyre pressures at 85psi always seemed somewhat daunting compared to the car tyre pressures I was used to and the rear inside tyre on the driver's side often gave trouble with a valve in the extension that didn't want to re-seat properly. In service stations, the compressed air supply by the petrol pumps was for cars and couldn't generate the pressures needed and those for trucks were with the diesel pumps that were often impossible to reach once you had filled with petrol, so it would have meant stopping specifically to check pressures and then, of course, tyres were too hot to be checked. On that basis I had bought a 12-volt TRUCKAIR compressor so we could do the job ourselves in camp when the tyres were cold.

Checking the tyres one day I found they were up to pressure but as the gauge was removed from the valve of one of the rear tyres on the driver's side it wouldn't shut off and the high-pressure air rushed out in a heart-stopping way down to 75psi before I persuaded the valve to re-seat. I set up the TRUCKAIR compressor for the first time but found its cable wouldn't reach the cigar lighter socket in Katie's dash or Boris's. I plugged in a newly acquired extension cable. It worked for five seconds then stopped. Each connection was wriggled to make better contact but to no avail. We plugged the compressor directly into Boris and it ran, confirming the extension was faulty. When I dismantled the plug I found a five-amp fuse had blown, so replaced it with an eight-amp fuse, the TRUCKAIR fired up for a few moments then cut out. I tried a fifteen-amp fuse without effect. I examined the plug in the end of the extension and found the fuse was fine but the spring supporting it had compressed, possibly even melted. No amount of tweezering would get it out and I thought I would have to manoeuvre Katie so I could get her rear wheel close to Boris when Gail inspirationally jammed some silver foil down onto the spring and the TRUCKAIR worked ever since.

On October 15 we woke to brilliant weather but even on my birthday shorts were banned as we were off for a leisurely lunch. Fréjus Port reminded me of Port Grimaud with modern flats surrounding a harbour packed with motor cruisers and expensive yachts but a lot of restaurants were closed and of those open none appealed. We are poor at selecting places to eat, often wandering around for hours looking for culinary inspiration.

At Fréjus Plage we returned to a restaurant where we had seen diners at tables with large blue plastic bowls brimming with discarded mussel shells. *'Les Moules Joyeux'* spilled out onto the pavement in true French style (but from the flags I think it might have been Belgian) and was nicely busy but not so cramped that we couldn't select a table with sufficient elbowroom for seafood scoffing.

We selected the *'Menu Marin'* and both chose *'Bouquet de Crevettes'* to start and *'Moules'* to follow. Gail had *'Provençal'* dressing for the moules and I selected the *'Aphrodisiaque'* with ginger, mushrooms and goodness knows what else. Davide who took our order raised his eyebrows and we laughed. Gail chose rosé wine; I had a celebratory Coca Cola, and we both had water presented in blue bottles matching the bowls. To round it off Gail enjoyed *'Île de* something' or other in *'Crème Anglais'* (custard) and I enjoyed two scoops of ice cream, with coffees to follow. As Michael Winner might say, the meal was historic. With the smell of crevettes lingering on our fingers we tried to walk off some of our gluttony in Fréjus centre buying postcards to send to the LGV driving instructors without whom none of the trip would have been possible. Later we received phone calls from family and friends for my birthday.

When we opened the curtains on Saturday October 18 it was raining, only the second time since the start of our trip. We felt the need to get out and decided to go to Le Muy, as it was the Mushroom Festival and *'Géant Omelette'* day. We drove straight for the centre and parked near the *'Old Piano'* restaurant; a policeman was on duty and all the uniformed school children were on their way home for lunch.

We walked down the street towards the church square; then we heard raised voices in front of us around a white car. Some men were remonstrating with the driver and passengers. We gave it no heed until I noticed that the driver and passengers were out and fighting had ensued. We quickly walked away and from a distance I saw one of the men extract a pickaxe from the boot of his car. As soon as it had started it died away. It seemed incongruous in the middle of a pretty but damp medieval village.

Yellow *'Géant Omelette'* signs took us to a school hall crowded with people sitting at long tables with bottles of water and a few stalls ranged around the perimeter. It was unclear what the system was or, who could go in. People continued to arrive and everyone seemed to know each other and where to sit. We withdrew disappointed not to have seen a *'Géant Omelette'*.

We were unstoppable tourists. We went to Mougins, one of the most wonderful places to visit on the Côte d'Azur. This superb medieval village set amongst pines, olives and cypress trees surrounded by forests offers panoramic views of the Baie de Cannes, the Lérins islands, Grasse and the *Préalpes*.

A few months earlier we had been for an evening dinner on a corporate junket with partners of the company I had recently retired from. As soon as we arrived in the village, we were seduced again by the charm of the narrow roads, bordered by colourful flowers and superb ancient houses. Picturesque doorways with each stone carefully restored, beautifully designed window-frames, delightful detail everywhere we looked and it was difficult to know where to point the camera.

The harmony of the colours and quality of light had seduced many artists and celebrities to stay in Mougins, including Picasso (who spent the last 15 years of his life there) and Winston Churchill and given rise to more than twenty art galleries and studios. On the village outskirts magnificent Mediterranean parks and gardens hid luxurious properties

Being only 15 minutes from Cannes, Mougins is a popular place to visit for dining with more than forty restaurants but we still couldn't choose the correct combination of value and a great culinary experience for lunch and left for Cannes.

We parked easily on the *Croisette*, walked round the harbour and the old town (*Le Suquet*) that becomes a quite magical dining experience at night and where we had eaten on previous trips but settled on a fish restaurant alongside the port where all the million-dollar yachts were moored tightly together.

'It seems a pity to pay all that money for the boat and the mooring just to be jammed up against your neighbour hearing everything they're doing and hoping they are not hearing everything you're doing,' I commented between mouthfuls of sole. 'If it was a campsite we wouldn't accept it.'

We walked past the film festival site that had been home to the stars the previous week but still hoped, like the other thousand tourists, that someone famous might have stayed on. We continued along almost the full 3 kilometres length of the famous *Croisette,* following the curve of the bay with the beach on one side and flowering gardens, lawns planted with palm trees and umbrella pines of the plush hotels on the roadside. We returned along the renowned Rue d'Antibes and its elegant fashion shops and abhorrently stared into one incongruously selling all manner of guns, bizarre knives, tazers, CS gas and other weapons of destruction.

That evening Gail tried to chew on something hard but found it was her own tooth, a large chip had come away.

The following morning she decided action was needed and our insurer recommended dentists in Fréjus and Cannes. We decided on Fréjus where we had the name of two dentists and armed ourselves with the phrase book that fortunately had an emergency section and for those more dentally challenged useful phrases to use if you cracked your false teeth.

In Fréjus Place Agricole we entered the old building of one of the dentists whose afternoon hours were shown as from 14:00 but at 14:30 were unable to raise an answer. A short walk away on Avenue de Verdon was the office of Dr Jean Martin whose brass plate showed that he would re-commence afternoon work at 3.00 p.m. As we walked away wondering where we could kill time, I saw a white-coated figure emerge. I raced across the street to ask if he was Dr Jean Martin and found he was but spoke no English. He took a sceptical pity on us and showed us up to his open-plan office that comprised desk and chairs at one end and treatment area at the other. Conversation in stilted French went slowly until I spotted a photo of a Jaguar convertible XK140 on the wall.

'This is yours?' I asked.

'Yes, I bought in Beaulieu, in the UK last month.'

I told him I had owned a 1955 fixed head XK140 coupe and all concern for Gail's tooth disappeared out of the window as we animatedly discussed the joy of ownership and became friends for life.

Responding to Gail's interruptive coughing, we stopped and negotiated a 55 euros price for the treatment, including the fascination of looking at instant X-Rays on his computer, then a local anaesthetic and a temporary (insisted on by the insurer) filling that wink, wink should last a long time. I took a surprise photo of Dr Martin with Gail as she sat in the chair a post-op dribble of pink mouthwash coming from her anaesthetised mouth.

What a storm in the night! Lightning, thunder, and heavy rain required me to get up and lower the satellite dish. We expected to wake up to a flood but didn't and the sun shone, enticing us to take a trip to Monte Carlo.

It took an easy hour of easy driving along the A8 and then a spectacular drive winding through many hills and tunnels. The final descent into Monaco was slow as it was single carriageway all the way but eventually we parked under the *'Casino'* and walked through the magnificent gardens to the stunning square where coffee at the Café de Paris was de rigueur and inviting in the sunshine. Many were doing the same, all hoping to see a famous face. The cost came in at 10 euros with tip but Gail thought her visit to the conveniences well worth it.

The harbour was full of yachts and surprisingly surrounded by a fairground but made a perfect lunch stop. We followed Rue Grimaldi and eventually came to the stairs leading up to the Palace and the old town and its narrow cobbled streets. It was beautiful and unexpected.

We were enamoured of Monte-Carlo and its sumptuous boutiques with the latest fashions, palatial luxury hotels, and chic restaurants. Even the shopping centre near the Casino, was beautifully appointed. Restaurant and café prices were inexpensive (except for the coffee at the Café de Paris) and lunch at the gourmet restaurant opposite the palace and overlooking the harbour could be had for 20 euros. Other restaurants were offering Italian meals for around 8-10 euros. The town or, country is amazing when you think of the actual acreage, the road infrastructure and tunnelling. It was a great day out and although not to everybody's taste, we say why not?

Friday October 24 started with the sun shining but it was chilly. We decided to explore the Massif des Maures and St-Tropez Peninsula. We set off west through Vidauban for La Garde-Freinet a mountain village. We chose a smaller road up (D558) because it looked exciting with many switchbacks. We found it was quiet and winding and provided a great contrast to the commercialism of the coast.

In La Garde-Freinet we parked in the centre, donned our fleeces and walked through the village, eventually arriving at an attractive stone property at one end. As we approached we saw a crucifix high on the hillside and then noticed a sign to *'Croix'* and *'Fort'*. We took the mountain path to the cross and after about ten to fifteen minutes we were at the summit with panoramic views to the sea and St-Tropez, the hills of Miremar, the plains of St-Clement, and the valley of Argens all the way to the Alps.

Back in the village square we stopped for coffee in the hotel lounge that was more like a pub. Later we saw in the estate agent's window that properties in this medieval village seemed to be in the £1 million range. Back at the car *'The Rough Guide'* told us Oxford professors now populated the village – I never realised academia was that well paid. They were well served by tempting food shops, local wines and a market twice a week. The villagers made a living from the cork and chestnut in the forests.

From La Garde-Freinet we headed to Gassin where we found almost perfectly restored properties perched high up on a rock and less than 4 kilometres from the sea creating a model village. According to *'The Rough Guide'* it was once an eighth-century Muslim stronghold. The views were wonderful, despite the wind and even the Post Office looked out across the vineyards, woods and mountains to the snowy Alps. The narrow winding streets between the stone houses were magical.

On our way back to the car we saw another road up the side of the village and on taking this we found three or four restaurants on a broad terrace, protected from the wind, with more magnificent views east. The prices were not exorbitant – this would make an ideal lunch venue or in summer a perfect location for an evening meal and we suspected many drove up from St-Tropez or Port Grimaud for this.

We headed to Ramatuelle, which we had been to before when searching for campsites but not to look at in any detail. This like Gassin is a walled city but it is more impressive as the walls tower high up from the hillside and the pink-roofed stone houses have not been restored to the same extent. We explored the narrow streets, which were like a maze and obviously built for shade and protection with only two or three gates.

We returned via Port Grimaud and turned in Ste-Maxime to take the D25 back over the mountains through a forest blackened and devastated by recent fires.

We were back at Katie and were beginning to use the word home and the sun continued to shine.

We started to think about moving, as the site would close on Friday, November 1 and decided to on October 30. Once we had taken the decision when to go we realised we had little time left and wanted to see many more places.

Whilst I have no intention to bore you with a simple travelogue Provence is a wonderful region to explore in the autumn.

Grasse had always been on Gail's list as she remembered us visiting on one of the previous trips to France and had memories of beautiful fields of flowers being harvested for perfume manufacture.

We used the *autoroute* as a quick means of getting into the area and had soon turned off and north up the D37 alongside Lac de St-Cassien, which could just as easily have been the English Lake District. We were then on the D562 round the hairpins and into Grasse. The day was grey, as was Grasse.

We parked next to the bus station and walked into the old city. It could possibly be atmospheric on a sunny day, but I doubted it. This is a medieval town that hasn't changed much, but appeared run down and poor. Apparently this was the case as almost all of the population had worked in the perfume industry for 300 years and was poorly paid.

We did the tourist walk using the map from the Tourist Bureau including the cathedral and other obligatory sites without too much inspiration. Eventually we reached the Fragonard parfumerie and joined a tour round the old factory conducted in French and after a while my brain wouldn't compute anymore so I wandered and then heard a guide speaking in English so grabbed Gail and we jumped tour parties, which also meant we finished quicker.

On leaving we knew we should eat but nothing looked attractive enough so we decided to eat what we had brought with us in the car and travel on up The Route Napoleon to St-Vallier-de-Thiey another fifteenth-century fortified village. The views demanded photos but we found nowhere to pull off the road. We left along the D5 to St-Cézaire-sur-Siagne and parked up. This is another pretty, artist-populated, medieval village clinging to cliffs overlooking a river. Hidden away at one end is a view right across the mountains to St-Tropez and the sea.

There were many caves in the area and having passed a sign for Les Grottes de St-Césaire we did an about turn and set off up a narrow lane eventually arriving at the car park about 2.15 p.m. with one other car. We wondered whether the caves were open. We approached the café where the man was optimistically putting out chairs and looked around for the entrance to the caves but couldn't find it.

'Où est Les Grottes?' I asked.

'Dans la maison.'

The entrance to the caves was inside the house to which the café was attached. In true French tradition they were closed for lunch and would reopen at 2.30 p.m. We had coffee and watched as cars started arriving then a minibus and it soon became crowded.

Our party was composed mainly of children and parents on holiday. The guide led us all into the caves (her talk was all in French though she spoke perfect English) and apparently was amusing (she played tunes on the stalactites) so in the interest of international relations we left a tip in her outstretched hand offered cordially as she blocked the narrow exit.

Intent on completing our visiting and with only two days remaining, we headed along the *autoroute* past Cannes and Antibes to Cagnes-sur-Mer (which is not on the sea), then inland up the D536 to St-Paul-de-Vence.

We enjoyed the view of the mountains and particularly the snow-covered Alps. The approach up the valley to St-Paul is beautiful and as we arrived a coach load of people was walking up the final hill confirming it must have something going for it. St-Paul sits proudly on the hillside like a fort on the mountaintop with spectacular views all the way to the Mediterranean.

Parking was difficult and we had to settle for the multi-storey built into the hillside.

As we walked towards the city walls we noted a café by the *boules* pitch was full and lively. The town was intricately interlaced with narrow cobbled streets between high-walled buildings. Quality art shops abounded, as did restaurants. We started to wander back then once again espied the café by the *boules* pitch and our thoughts turned to food. We wandered in; it was packed with tables close to one another and just how I remembered the Paris bistros. The waiter effortlessly gathered us up and we were soon sitting down at a table, the paper cloth laid, glasses provided along with cutlery before we knew what had happened.

No menu was provided and we joked about how the last time this happened I ended up with steak tartare and Gail with pig's ear and tail. After a short interval the waiter who appeared to be serving about fifteen tables of four or more, put his hand on my shoulder

'Salade niçoise, entrecôte et frites, saucisse,' and then he left.

We assumed he has described the menu, rather than you getting all three and sure enough he was back and we correctly selected salad for Gail and steak *pour moi* with wine (for Gail) and water.

The waiter was managing fifty tasks at one time but never missed anything. The meal was great, the steak enormous and we enjoyed every moment.

As we emerged, the sun had decided to shine and we bought some postcards then wandered down the hill to get the ultimate photos of St-Paul on the mountain, the sea and the white-capped Alps – a unique and exceptional place.

The tour continued through Vence, just as interesting, but we didn't stop and continued along the mountain route D2210 through Torrettes-s-Loup down into Pont-s-Loup, and through le Bar-s-Loupe with great views down into the Gorges and beyond to the coast. Soon we were in Grasse again (it hadn't improved since the day before) and on towards Cannes on the N85 to pick up the A8 *autoroute* back home. As we approached 'our' local mountains the 'Maures' the sun broke through and we wondered whether Katie had been in the sun all day and hoped that tomorrow would bring a rise in temperature. It was some comfort knowing we were ten degrees hotter than the UK but we wanted continuous sun.

Our last full day at Le Muy had arrived and it was time to prepare for departure. It was grey and raining with puddles everywhere. As we ate breakfast and I looked sleepily out of the windows, I saw that the electricity cable was under water and at that moment the electric cut out. To save going out in the rain and so Gail could use the hairdryer I started the generator. It started more slowly owing to the damp conditions and had my every sympathy.

After getting the coiffure done and allowing the generator to run down with no load I decided to go out and put the trip back. By now we had a deluge. I donned sneakers, shorts, T-shirt, precautionary yellow gloves and hoisted an umbrella. Gail was thoroughly amused by this apparition and rather than expounding sympathies tried to hold herself steady enough to take photos through the window.

A final batch of cards were written and posted at the campsite office and the bill paid. The weather relented a little bit and suitably attired we undertook our last trip to 'Ed's' for top up shopping.

There was almost a hint of sunshine on our return so we took some photos of the site (trying to hide the puddles). We put the warning signs and triangular reflectors on Boris ready for the trip tomorrow and then took a final walk round the site. It really is enormous, entirely natural and deserted apart from us and one other occupant. We wondered what tomorrow would bring it was a month since we had travelled in Katie.

Vauvert

'Why are you taking photos as we drive along?' I asked.

'Those mountains are named Victor, the same as my dad.'

We had set off west along the A8 heading for Nîmes, on the border between Provence and Languedoc. After our long stay at Le Muy it had seemed strange getting Boris hitched onto the back of Katie but everything had gone smoothly and it was a beautiful clear dry day. The road was quiet and I felt pleased to be back in the cab covering new ground and we were relaxed.

Péages, and there had been many throughout our journey, were no longer obstacles of fear because of their narrow dimensions and we took them in our stride. As another one loomed we chose one of the only two lanes open.

'*Bonjour, c'est combien?*' I asked of the vacant young girl in the glass booth but received not a word of reply.

'Classe 3, 18.00 euros' came up on the indicator below me and I relayed the amount to Gail for the money.

'*Vous avez une voiture aussi?*' said Mademoiselle Glass Booth suddenly waking to spy Boris lurking round the back of Katie.

Now nobody had bothered about Boris before, since he was clearly attached to Katie or, Katie's bulk obscured him from view.

'She's asking about Boris, shit, she wants another six euros for him.'

'*Ce n'est pas une voiture, c'est une remorque,*' I explained.

'What's happening?' Gail asks, unable to hear the exchange.

'I've just told her Boris isn't a car but a trailer.'

'*Non, c'est une voiture, six euros.*'

'She's getting insistent,' I relayed across the cab 'and there's a lot of traffic building up behind us. I'd better put the flashers on.'

The 'it is,' 'is not,' exchanges continued as the mêlée of traffic now spread out behind tried to untangle itself and move to the only other available tollbooth.

'*Où est votre chef?*' I tried.

The partition on the booth was slammed shut and Miss Glass Booth was on the phone.

'What's happening?'

'I've asked her to get her chief – that is what '*chef*' means, isn't it, otherwise we're going to get lunch served?'

Mademoiselle Chef arrived, supported her colleague and we went through it all again. I told her Boris had little red triangles on because he is acting as a trailer. If he is a car, where is the chauffeur, and why is the motor not running? I thrust a paper we have explaining such things in French.

'What's happening, the traffic looks bad behind us?'

'This one's no different – they still won't accept Boris as a trailer.'

'Perhaps we should just pay and go?'

'Not bloody likely. According to the tariff on the booth they're trying to charge us more than the maximum you'd pay for a forty-tonne truck – how can that be right?'

'Why don't we offer to pay as a truck?'

I proffered another 5 euros and they took it, closed the window again but nothing changed and the barrier stayed firmly in place. We argued some more and she mentioned the police and I said good and closed my window.

'What's happening?'

'She's dialling the police.'

'Oh no! Now look what you've done. We should have just paid. What about the A-frame? They're going to be really mad when they see this traffic chaos.'

'They probably won't even come,' I tried to pacify.

We could see the flashing blue lights half a mile away and the two cars and three burly gendarmes fully tooled up were there in less than a minute, tyre squealing to a halt.

'*Bonjour, Monsieur Gendarme, il y a un petit problème.*'

To be super friendly I got down from the cab and in my best Franglais went over the previous ground of red triangles, no driver, no motor running and firmly attached to Katie LIKE A TRAILER and what's more we'd never had to pay at any other toll booth throughout France and why should we start paying more than 40-tonne trucks?

'Eez different in de South France,' one of them said.

'*Bienvenue au Sud de France*,' I incautiously responded and he nearly laughed.

Mr Senior Gendarme said he didn't really know and hand on revolver suggested it might be best to pay the extra. I handed over the outstanding 1 euro.

'*Un billet, s'il vous plait*,' I said.

'What's happening?'

'I've asked them for a receipt but their machine can't do it because the barrier will only go up once for us yet they've charged for two vehicles so it will look like they've been cheating. The gendarme is asking Miss Glass Booth why she can't do it. There's a big debate going on. This is fun.'

Eventually, the gendarme passed a handwritten note to me and wished us a polite '*Bonne journée.*'

I almost felt vindicated.

'That was horrible,' said Gail.

'But did you notice they didn't arrest us for the A-frame?' I responded.

A few miles farther on we passed smoothly through the next tollbooth as a 'Classe 3' so they obviously hadn't radioed ahead, but then again perhaps they had.

Even with the delay we approached Nîmes early and were soon on the Nîmes Périphérique. We located the D13 and then '*Domaine de la Bastide Camping*' and parked in the spacious area in front that seemed to double up as a bus terminus.

All was quiet but we should have known, as it was lunchtime. The reception was empty so we had a walk around the site. It was level, well planted without creating turning problems, but had a lot of static caravans. We couldn't get excited or comfortable with it and decided to continue to the next site. If we found it to be no better we would return. Indecision often prevailed at a new site – we would try to see ourselves settled in the new environment – would it be what we wanted, were the facilities good, how would the neighbours be, did they have a tribe of kids or barking dogs, where did the sun rise and set – just like buying a house but more often. When you were on the road you were in control but as soon as you pitched you gave up control and other factors took over. *(Three years later we stayed at the site and found it excellent.)*

We drove down the D135 with open fields on each side, vineyards and a few barns and on the right near the end of the road just before it joined the D572 to Vauvert we found a huge sign announcing *'Camping Les Tourrades'*. A wide dirt road provided access with white-painted buildings either side leading to the entrance at the end. Russet-brown chickens clucked behind wire mesh, an Alsatian guard dog lolloped up and down in a pen behind a low wall and a little black dog ran out barking a greeting at us chased by a lady who stopped suddenly at the sight of Katie (I don't think she could see Boris). She told us to have a look around the site but thought that if we wanted a pitch with electricity the trees would make access difficult. The long, narrow, grassy pitches were regimented and separated by high hedges with trees planted in them. We wandered round and decided to stay – as it had an easily accessible dumpsite and I thought we could wriggle onto pitch 142. As we drove in, the lady watched in awe as Boris followed Katie.

We detached Boris and reversed onto the pitch, set up straightforwardly, and welcomed the water tap and 10 amps electric supply on the grassy pitch. There were few other campers. We had driven 142 miles and as it was only 2.00 p.m. I suggested we went to Nîmes in Boris and check it out – was it worth staying?

In Nîmes we luckily found on-street parking for two hours and did a tour of the old town that has some of the most extensive Roman remains in Europe and is equally famous for denim (de Nîmes) the material sent out to clothe workers in the southern United States. We eventually ended up at The Roman Arena, a spectacular two-storied and galleried 20,000-capacity building; apparently one of the best preserved anywhere and currently a premier centre for bullfighting but we were too late for the guided tour. Peering through the gateway it was easy to imagine epic Hollywood-style productions of chariot racing and gladiatorial bloodbaths. A sign for *'Vomitoires'* in the arena's entrance to the labyrinth of corridors provided five minutes of game-show guessing as to the possible meaning that was unresolved by recourse to our mini French-English dictionary. Later with some *Googling* and *Babelfishing* I found out they were the openings in the steps from the network of subterranean passages and doors under the steps allowing spectators to enter and exit these ancient theatres. Furthermore the spectators were *'out-spit'* from the amphitheatre in such a way that the entire crowd could withdraw itself, almost at the same time, *'without the least obstruction and the least embarrassment.'* It was calculated that a coliseum able to contain 90.000 spectators should have enough *vomitoires* and staircases that its public could disperse in less than five minutes. Wembley stadium are you reading this?

The next day started out misty and damp. As I took a walk along the campsite road in a relaxed mood, I noticed an English couple packing up their caravan. I said hello and stopped to chat expecting to exchange the usual pleasantries about where they had come from and were going to.

'We were driving in Spain when a car drew alongside, the back window went down and a gun appeared. They shot at our caravan tyre puncturing it so that we were forced to stop. Fortunately an English truck driver saw what happened and pulled in to protect us, otherwise we think we would have been robbed.'

I listened aghast. Could it be true? They recounted tales of attacks on other motorhomers and caravanners. It sounded too far-fetched to comprehend but it was clear from the strained expressions on their faces that it had happened to them. They couldn't hit the road out of Spain fast enough.

The abrupt jolt to my pleasant 'good morning' had left me worried and unsure what to do. Spain began to sound like some wild frontier land. Should I even tell Gail? I thought it over as I wandered the rest of the campsite.

I decided to discuss it with her but we didn't arrive at any conclusion other than we should take advantage of where we were. France seemed a safe haven.

'What a stupid bloody map,' I moaned.

Trying to look at the Nîmes Tourist Office's tour map and read the upside down information key on the other side required constant page turning that proved impossible whilst holding an umbrella in windy and rainy conditions. Nevertheless the *'Jardin de la Fontaine'*, France's first public garden, was beautiful with fountains, lakes, canals and winding paths through grottoes up to the Roman tower at the top. The view from the tower shown in the brochure was spectacular. Why was it that even if we started out in perfect sunshine by the time we reached a viewpoint clouds or mist had descended? We could just make out the Temple of Diana below. It looked neglected for such an important historical monument and the skate boarders whizzing through the surrounding gardens a little incongruous.

Back in town, The Roman *'Maison Carre'* built in 5 AD was an impressive beautifully proportioned and columned history book temple and little imagination was required to see it as the centre of Roman city life. I pictured all those centuries ago the edicts being read out to the populace from the top of the forum steps and then wondered how the Romans coped with their togas and sandals on similar rain-swept days. Had the Roman equivalent of 'Wellingtons' been invented? (In researching this later I found that contrary to my idea Romans wore sandals inside and shoes outside, it was impolite for a *'free man'* to wear shoes inside so when he walked he wore shoes and his slave carried his sandals). Across the square and in complete contrast was Norman Foster's modern glass *'Carre d'Art'* a giant pyramidal greenhouse of a building so we enthusiastically dashed in, even though we were wearing shoes, to a light-flooded welcome respite from the rain only to find most of the exhibits closed.

We wandered through the maze of narrow streets enjoying the history of the old city but disappointed the dampening weather prevented us enjoying a coffee outside in one of the many squares that popped up unexpectedly. We learnt that Dr Jean Nicot born into the city has the dubious honour of introducing tobacco into France from Portugal in 1560 and giving his name to the drug.

'C'est un grand camping-car.' Vous avez beaucoup des enfants?' asked a passing French camper.

'Non. C'est à deux,' I replied.

He continued to stare in amazement that we should be so indulgent as to have such a huge vehicle just for the two of us. He stood transfixed by Katie as I made an attempt to explain a 6.8 litres Ford V10 petrol engine and other mechanical features in French.

The day dawned with better weather and our enthusiasm for sightseeing was re-kindled.

The name of Aigues-Mortes (dead waters) held a certain fascination although it was hard to imagine why but it was less than 15 minutes by road along the N572 then the D62 a flat Camargue road floating on water as far as the eye could see. As the massive fortress walls and rounded towers of Aigues-Mortes came into view, we turned left and parked beyond the many busy sandy *boule* pitches broken up by wooden planks. We stopped to take in this slice of French life and bonhomie. The six players on the pitch nearest began their game by throwing the small bright green plastic target ball. It was the turn of the one in jeans and yellow and blue striped T-shirt that barely covered a commodious stomach. I wondered how he was going to bend down to pick up his *boule* when in one moment he had let slip a chain in his left hand that descended to the *boule* attracted it with the magnet on the end and he deftly flicked it back up into his hand. He stared intently at the sandy pitch, swung his arm back and let fly with a boule that landed a foot from the green marker. The player dressed in chequered cloth cap and winter brown sweater with patterned snowflakes and brown corduroy pants did the same, landing a foot the other side and the game was underway. As each took his turn to throw his *boule* there was much positioning advice and gesticulation from the two team-mates crouched over the *boules* already thrown. When all had thrown, the group of six gathered in deep concentration to examine the results and it was clear this was not an easy one to decide. Mr Commodious Stomach was quickest off the mark with much energetic pointing and arm waving but the younger Mr White Jeans, Adidas blouson and sunglasses perched on his head, soon joined in. The elder but inhibited player in his black trainers, checked shirt and brown gilete hung back and looked out of his depth. The young and impetuous car-dealer type in blue jeans and short blouson was soon in amongst it whilst Mr Brown Ski Sweater ruefully scratched his head. Would this ever be decided? But then spoke the elder and revered player in blue tennis shirt and fawn trousers and silence prevailed when he trumped them all by producing a tape measure from his pocket. With everybody staring acutely at him he knelt to measure the distances between the *boules*. Knowledgeable onlookers seated at the edge of the pitch nudged their neighbours and re-enacted his tape measuring action. He straightened up and with a finger at the crucial point on the tape measure pronounced to the others now standing in an arc around him. There was still no clear result. Shoulders drooped disbelievingly and they looked to each other for salvation at which point we departed. Is it wrong to say that *boules* is merely marbles for adults?

Aigues-Mortes initially had that wow factor. The guidebook said Louis IX built it as a defensive port in the 13C for his departure on the Seventh Crusade. It is surrounded by a rectangle of intact solid honey-coloured fortress walls, towers and battlements like a picture book crusader's castle. Colourful arrays of flowers in the pavement gardens adorned the entrance gates. We walked in through Port de la Gardette and onto the main street Grand Rue Jean Jaures that had touristy shops but was characterful and at the end the town square was packed with expectant colourful restaurant tables and chairs. It all looked smart and the menus good value. The church had pleasing stonework but unappealing modern stained glass windows. We walked through the town, found we couldn't climb the walls, and so exited through them to the other side. Ahead of us the flat Camargue stretched into the distance with mountains of extracted salt glistening white in the sunshine.

Close to Aigues-Mortes and on The Mediterranean is Port-Camargue a modern marina without the class of St-Raphaël or Port Grimaud but pleasant nonetheless. Being a holiday many French were taking the opportunity to walk their dogs.

We continued to be amazed how much dog poop there was everywhere we had been in France. It was impossible to take a walk without keeping your head down and in so doing miss many of the sights and views. We yelled to each other (poop warning) as we saw the possibility of them stepping into a pile. We had seen one sign for no defecation; it had been on a *boules* pitch. We had never seen anyone poop scooping.

After Port Camargue we went into Le-Grau-du-Roi. The sea-front candy stalls reminded me of Blackpool but as we wandered round the front then along the *Quai* it changed dramatically, becoming a working fishing port with the river or Canal du Rhone running right through the town with small colourful fishing boats double-moored to the quayside on which restaurants and shops abounded. Farther on were the more industrial areas of the port with larger dry-docked boats, their rust-encrusted hulls contrasting sharply with an azure-blue sky and large white fluffy clouds.

As the sun set, we walked back through town – each street was lined with tacky souvenir shops but one stall, selling pastries and ice creams, attracted a long queue of children and adults all buying a cone-shaped bag of what looked like giant chips but weren't. We joined in and bought a bag of deep fried *'Churros'* – delicious strips of fried dough dipped in sugar or chocolate sauce – we were kids again. We wolfed into them as we wandered back to our starting point and finished off the last few sitting on the seaside wall – it could just as easily have been Southend. Thank goodness we only had the one bag – we felt stuffed.

Palavas-les-Flots appeared from the map to be attractively floating on a thin strip of land between the sea, a canal and lake. It was also the location of an award-winning *aire de service* for motorhomes. Despite the beauty of a Mediterranean sky, the approach from Carnon Plage was dispiriting – the road was either one house from the sea on the left, or a lake on the right, and every house was a stereotypical Blackpool B&B but without the vacancy signs in the windows.

Eventually we arrived in Palavas-les-Flots and followed signs to the *aire*. This took us alongside the canal to a strip of land dividing the canal from the marina. It was basically a tarmac car park but well-kept and provided electricity; a barrier with an attendant and an external dump station. There were many white motorhomes basking in the heat of the day, their owners ministering to them and as we glanced beyond the canal to the Camargue, we saw the flash of pink from around thirty flamingos' heads down in the reed beds.

We walked alongside the canal back into town. It was jammed with fishing boats jostling and bobbing against the quayside, their rigging singing in the light breeze. On many, the polo-necked and rubber-aproned crew were busy selling off their catch to eager purchasers leaning over with blue plastic bags at the ready. Cafes, bars and restaurants occupied the ground floors and spilled out of the old buildings lining each side of the quayside. The sun had brought out the crowds and the cafes and bars were busy and we were only too pleased to join in the Venetian scene and have coffee.

Our next stop, Sète, was a lot farther south, so we hopped onto the motorway. Following *'Centre Ville'* signs we were soon back at the Rhone Canal (yes I know it entered the sea at Grau-du-Roi but that was a branch). Sète has apparently been an important port for 300 years. We were delighted to discover a repeat of the scene we found at Palavas-les-Flots with more macho and workmanlike quayside trawlers and fishing boats fronting the busy and extensive waterside restaurants, cafes and bars that made the quest for a seafood lunch something along the lines of kids in a candy store. A post-prandial walk through the town led us to a steep hill up to the main attractions: the Cimetiére Marin – the sailors' cemetery, Cimetiére le Py and a view of the Bassin de Thau, the inland lake full of mussels and oyster beds. A hundred yards up with no sign of the aforementioned and an extravaganza of a seafood lunch weighing heavy, the appeal of viewing gravestones rapidly diminished so we turned around, went back, and headed for home.

With days of continuous sunshine we chose to do the full tourist bit and headed Boris northwest up the A9 for about 45 minutes to The Pont du Gard. A vast, but empty sea of white stonechip car parks, greeted us. The modern aircraft hangar that doubled as a reception and media centre had a bar-restaurant and the staff were pleased to get the steamer going for the first coffees of the day. We walked the new concrete path dodging the busy JCBs and blue-overalled workmen laying even more walkways to the Pont.

It was us and a whole load of first-century Roman engineering stonemason magnificence built to bring fresh water 50 kilometres from the hills to the city of Nîmes. We walked across the three tiers of arches spanning the River Gard searching the stonework for the Roman builders' markings – 'front side left number 3' – but in Latin, of course. 'FR S III' then up the opposite hillside to the *'Point of View.'* We dwelled on just how with no modern machinery, computers or the like, such a perfect structure could be created so long ago. A Japanese family joined us and I took a photo for them and they reciprocated with a bow so I bowed and they bowed again so I thought I'd better stop or we might be there all day.

Afterwards, we headed northeast to Uzes about 15 minutes drive away. At a pavement restaurant we enjoyed lunch then explored this lovely medieval town with its towers, castellated walls and narrow streets that support bull running in the summer. The Castle of le Duche, in the centre of town, has housed the same family for a thousand years and The Tour Fenestrelle looked like a replica of the leaning tower of Pisa.

On the way back we diverted from the main D981 along a country road through Vers to the small hilltop village of Castillon-du-Gard for no better reason than it looked attractive on the hillside. It was pretty inside as well with a charming square full of umbrella-shaded tables and chairs outside the restaurant. We walked round the walls of the delightful church stepping carefully around the men playing *boules* and looked back down on the road winding its way across the flat countryside and from whence we had come.

Domestic chores at the campsite were a breeze. The campground bar-restaurant seemed to be the meeting point for the locals and they watched with interest as we drove slowly past to the dump negotiating the overhanging trees, Gail giving instructions into the walkie-talkie. With a warming sun, emptying the tanks was almost pleasurable and with our new hose we were able to give the holding tank a really good rinse. With the water supply on the pitch we could fill our tank whenever we wanted and spend time thoroughly cleaning Katie and Boris.

Washing was done in the camp laundrette that had two washing machines but no dryer so we used our clothesline. Enthused by this domesticity, I rushed off to Vauvert and bought some coloured pegs in a snazzy basket and then later what my mother would have called a *'clothes-horse'* so that we could dry inside should we have inclement weather.

We thought about moving on and drove to a couple of campsites near Montpelier, one pleasant the other we were put off by the guardian announcing, as we entered his office, that he was off to lunch and would be back in a couple of hours.

Montpelier turned out to be a lively city full of young students. We had a good lunch in the crowded shopping centre and toured the pleasant pedestrianized streets of the old town and although we felt we had done it, I sensed Montpelier had a lot more to offer than we had time to discover.

As time passed, the weather became less certain and rain more frequent although since the holiday guides said there would be 360 days of sunshine, by definition the rain had already fulfilled its quota. Despite the attraction of World

Cup rugby on TV and gloomy conditions, Gail was anxious to get out, so we set off for Gallargues-le-Monteux, a small sleepy rural village with approximately 3,000 inhabitants in Languedoc-Roussillon.

On the map was a *'Pont Romaine'* and according to The Rough Guide the Roman Via Domitia leading from Gallargues to the Pont over the river Vidourle. On the west bank of the river is said to be the old cobbled way climbing the slopes to the former Roman settlement of Ambrussum, a fortified staging post on the road. I had formed a picture in my mind and looked forward to seeing the reality.

We found Gallargues and with such a large write-up in the travel guide expected signs to this famous ancient monument but we didn't see any. Looking again at the map, we decided on an alternative approach from Lunel going towards Sommieres taking a turn onto the D110E, towards Villetelle. All this was close to the A9 *péage* with confusing twists and switchbacks so we took great care to avoid ending up on the motorway. The D110E was a narrow country road and as we approached the railway bridge we spotted a track off to the right but still no signs. The cart track wandered past vineyards, car scrapyards and what appeared to be a gypsy encampment, then into a nothingness of sand-dune-like grassy mounds and a dusty road. A fork in the road showed a tiny arrow painted on a concrete telegraph post: *'Ambrussum.'* Half way down the new track we came across a fork to the right and another faded, barely discernable, *'Ambrussum'* sign.

Eventually we reached a hedged-off space by a river, an architect's sign from the council of Lunel and the Pont – covered in scaffolding.

It seemed natural to wander up the hill. Close to the top was a cobbled street with ruts in it where the ancient carts must have run. We took a little path and started discovering ramparts, walls and building foundations but almost all of the stones had fallen from the walls. We stood entranced by the remains of what was a complete Roman fortified settlement guarding the route to the town of Gallargues that had become visible in the distance.

I learnt that we were standing on the oldest Roman road in Gaul. It linked Rome in Italy to Cadiz in Spain. This section had allowed Rome to administer the whole of southern Gaul, distributing agricultural land to Roman colonists and building new towns. A 'crossroads life' developed all along the road, where it linked with neighbouring towns and boosted the local economy.

The road was neglected for centuries, and a modern road hides the original materials in many places, but whole sections of the foundations and engineering works, such as bridges and mileposts, can still be seen. The Via Domitia is one of the major landmarks of France that has left its mark and shaped the landscape of the area forever. It seemed a shame it was hidden away and uncared for. Close by we could see traffic speeding along the modern international equivalents, the *autoroute* and parallel rail line the construction of which had covered over much of this magic history.

We left reluctantly and headed northwest to Sommieres, a gated medieval town with narrow alleyways. It was late and in the fading light the small shops in the solid stone buildings with cobbled courtyards were all brightly lit, resembling a Christmas card scene. Children were playing street games and only the horse-drawn stagecoach with a red-coated driver was missing. We walked through the narrow and dark alleys and up the hill to the chateau that was still occupied by the owning family, along the walls and through the gardens and then down again to one of the gated entrances in the town walls. Tempted by an amazing array of confectionery we had coffee and cake in what I would have called a milk bar in the old days.

On Tuesday November 11, Armistice Day, an English couple Peter and Jenny arrived in their car and caravan and parked behind us. They had come from Italy and were heading to Spain. Peter was taking a year out after retiring from the police force and Jenny had extended leave from Barclays Bank. Ben, a well-behaved golden retriever had accompanied them and had been popular, particularly in Italy. They were the first Brits we had had chance to get to know and it seemed a shame that we had planned to leave the next day.

After lunch and in preparation for leaving we dumped the tanks then drove off in Katie to fill up with propane (GPL) at the local garage. We squeezed into the bay next to the pump. After fitting the adaptor we had brought with us from the UK to the LPG tank we stood like a pair of idiots trying to work out from the French instructions and pictures the method of filling. A car pulled in behind us.

'Pouvez-vous m'aider?' I enquired of the driver.

He showed us just how easy it was and stood back nonplussed by the small amount we needed for such a big vehicle. I explained it was only for the *'chauffage'*.

The continued sunshine induced us to take a trip to maximise our stay. I had seen an intricately-painted sign to a local village called *'Le Callar'* so we decided to explore and found a pretty but sleepy place that we walked around in ten minutes.

With some time available and the sun still shining we decided on further exploration and headed for St-Gilles, a small town to the east of Vauvert on the medieval pilgrims' route to Santiago de Compostela in Northern Spain. In contrast to Le Callar, St-Gilles was enjoying the holiday, and was crowded. We checked out the port area on The Canal du Midi then wandered into the medieval walled town that was devoid of tourists and a bit run-down. We stopped for a coffee on the high street in a tiny step-down-into café with the locals and sat alongside the humming ice-cream refrigerator.

Instead of returning back to Vauvert, we headed farther into The Camargue hoping to get to the coastal resort of Stes-Maries-de-la-Mer where the Romanies hold their May festival in honour of their legendary dark-skinned patron 'black' Ste Sarah, the servant of Mary Jacob, Jesus' aunt who with Mary Salome, mother of apostles James and John, Mary Magdalene, Lazarus and his sister Martha beached there after fleeing persecution in Palestine. Festivals involve *'guardiens'* riding white horses into the sea along with the effigies first of Sarah and on the following day Mary Jacob and Salome both sitting in a little wooden boat. The road was exactly as we pictured the delta land of the Camargue – swampy water on each side of the drained area, screens of bamboo and reeds, but few trees. We had occasional glimpses of the famed white horses and horned black bulls all silhouetted with the setting sun just starting to redden the sky. As we drove south down the D37 then the D570, the traffic coming from the opposite direction was solid and included large numbers of cars towing horse trailers. There had obviously been a holiday event on the coast. The traffic increased and with the sun lowered farther in the sky we turned west along the D38 to Aigues-Mortes, hoping to take the little D779 along the Canal des Capettes directly to Vauvert. As we made the turn the sign to Vauvert had been crossed out suggesting the road was closed so we returned via Aigues-Mortes.

The next day we would move on.

Colombiers & Bizanet

'Let's not make any mistakes,' I pleaded to Gail.

I had brought Boris and Katie out of the Vauvert campsite into the entrance road so we could hook them up. Word circulated and a large crowd of aged men were sitting on the wall waiting for the show – where had they all come from – had messages been sent out to the village? Every move we made was being scrutinised, analysed and discussed with much gesticulation of the hands, pointing of walking sticks, and removal of hats and scratching of heads. We were as efficient as possible and I felt we should take a bow at the end but substituted a cheery wave instead and received many in return.

We followed the N572 then the N313 up to the A9 motorway then down past Montpelier and Beziers leaving at junction 36 onto the D64 then north onto the D11 until we found the left turn onto the D162 towards Colombiers.

We were on a country lane with grassed fields each side and proceeded as slowly as possible, in our usual style, to find the turn. A narrow humped bridge over the Canal du Midi appeared ahead but we thankfully spotted our turn just before it and down to the left. It narrowed and we took a right, then a left running between fields with cattle, horses and goats to arrive at the two crumbling stone pillars of the entrance to Les Peupliers camping on our left. It had that out-of-season neglected look; the many trees having shed their foliages and laid a deep rust-coloured leaf carpet on the lane and into the campsite.

The owner appeared promptly from his chalet bungalow and we tried to decide if we could drive in and how. We unhitched Boris in the road and I decided the only way was by reversing. With Gail on the walkie-talkie, the owner at the front making the usual wheel turning signs and me ignoring them both, we were in, parked by the entry gate, and hooked up. As far as I could determine we were the only residents.

We took a walk to the village that was immediately adjacent to the bridge by the Canal du Midi and had a small modern inland 'port' full of hire boats. From the many guidebook pictures it was how I wanted the Canal du Midi to look but it seemed even more special with the autumn colours and the *Chateau* at the water's edge. Probably a hive of activity during the summer season, it was a haven of peace and quiet with only a few local teenagers hanging about and surprised to see strangers.

'Where are you?' asked my mother during our weekly mobile phone conversation.

'Near Béziers.'

'Oh, that's twinned with Stockport. The children come here on exchange visits.'

Despite the warning, we drove to Béziers; its castle-like cathedral standing grandly on the hill as we approached from the surrounding plain of vineyards. I thought the children from Béziers would feel hard done-by with Stockport viaduct and the dregs of the River Mersey. I mentioned the twinning to the man in the tourist office but he seemed to be in denial of the relation. We did the suggested tour including the Cathedral St-Nazaire which had the gruesome reputation of being rebuilt on the site of the crusaders' massacre of seven thousand good people of Béziers who refused to hand over about twenty Cathars. Today, it provides magnificent views out over the river and plain. In the Place de la Revolution many had died resisting Napoleon III's coup d'etat in 1851 and even in the 1970s ugly events occurred as Occitan activists helped to organize the militant local vine-growers when blood was shed in violent confrontations with the authorities over the importation of cheap foreign wines and the low prices paid for the essentially poor-grade local product. Things are now more settled and any anger is taken out on the Rugby field.

I tired of all the history and it had started raining so we headed for the main concourse (Allees Paul Riquet) and at its junction with Place de Jean Jaures we settled for a coffee and an escape from the rain. As we sat and recuperated I noted the man at the next table had a Lidl's carrier bag so I asked him where the supermarket was.

The town was a nightmare for traffic but we eventually found a Lidl on one of the roads that led out of town but had no car park so I dropped Gail off then parked lower down the hill. The area felt run-down, it was raining and the light was fading; perhaps it should be twinned with Stockport after all. As I walked back I had to pass a vicious, brown snarling dog of the bull terrier kind that I was sure would be illegal in the UK with a short squashed skull, muscular chest between wide-set legs, smooth stocky body topped by a pert twitching tail. It had taken an implacable stance across the pavement blocking it to all but the brave or foolhardy but there wasn't anyone else around to test it, only me. I took a wide berth into the road and risked the oncoming rush-hour traffic. Relieved to have passed unscathed I then heard a car alarm that sounded like Boris's. I was worried someone had broken into the tourist's car and had to go back. Once more that dog eyed me up, it sensed my fear, and growled menacingly. I took an even wider detour into the road, car horns sounding. It was Boris's alarm – I hadn't closed the rear door completely. Having gone down the hill I had to return once more back up the hill to face that snarling, slavering, pink-gummed, sabre-toothed beast, hoping it wouldn't get third time lucky. I ran across two streams of traffic to the other side of the road as it seemed the least-worst alternative. We did the shopping and as we left I warned Gail of the ferocious animal we would encounter down the hill between Boris and us. It was nowhere to be seen so she thought I was exaggerating as usual. We headed home and with good fortune found our way out of Béziers but the weather was no better.

The next day The Canal-du-Midi was bathed in the morning sunshine, autumnal russet leaves carpeted its banks and the few remaining on the trees finally let go and fluttered down to gently float in the water, all wonderfully photogenic. We drove to Valras then through Serignan and onto Agde and Cap d'Agde.

Agde is situated on a river and crowded onto the east side. We did a town walk amongst the black walled buildings arising from the volcanic stone of the Mont St-Loup quarries and came across a fascinating area with a series of murals depicting buildings and people entering them and had fun taking stupid photographs alongside them. It was hard to distinguish between what was reality and painting, which I guess was the purpose.

We spotted an art shop down one of the lanes and debated with the artist the merits of a picture suitable for our lounge wall. I did the man thing of saying if that is what Gail wanted then she should go ahead. The usual excuse when you are abroad of transporting it no longer applied as Katie had loads of room. Gail decided against – I understand other women may not behave in this way.

We went onto Cap Agde on the Med. major resort housing up to 20,000 naturists but saw none.

Our next stop was Pezenas situated northeast of Béziers. We somehow missed the views *across to rice fields and shallow lagoons, hazy with heat and dotted with pink flamingos.* Following the tourist office's guide we saw the town's *14-17-century mansions part of Armand de Bourbon's plan to create a second Versailles.* Many were restored or, being restored, and we peered into cave-like shells with cement mixers and dusty workmen and magnificent fireplaces still clinging precariously to the first- and second-floor walls.

We wanted to visit the major tourist attraction of Carcassonne but with only two other campers on the site and the falling autumn leaves laying a carpet on Katie's roof, I also wanted to make a move. We could investigate a campsite at Bizanet on the way.

'Look Peter, Jenny and Ben the dog,' I exclaimed.

As we pulled into the Bizanet campground our neighbours at Vauvert were there so we drove straight over to their pitch. Willem the campsite owner steamed over to remonstrate about the need to go into Reception first and the other twenty rules we had broken but after explanations all was well and we agreed that we would bring Katie over the next day.

We went onto Carcassonne, deciding on a cross-country route rather than the motorway. The road through Bizanet village looked best but with no signs we were soon lost in the narrow back streets. Eventually we found the left turn we were seeking and headed along the country lane to join the D613 towards Lagrasse.

On the road we saw a monastery, as well as stunning scenery. We were delighted with the feeling of being in rural France with vineyards each side of deserted country roads.

The road through Lagrasse, *one of the most beautiful villages in France* (this became something of a joke when we realised how frequently this accolade appeared to have been awarded) and into the Gorge then Servies-en-Val and Pradelles-en-Val was spectacular and we were overjoyed to have gone that way.

Eventually we reached Carcassonne and drove into town through the back roads joining the busy main road from Montpelier, then across the Pont Neuf to get to the tourist office in Square Gambetta. The modern town was totally separated from the old part so after obtaining the information we drove back to access la Cité perched on top of the hill and spent considerable time exploring the narrow streets inside the magnificent double walls and turrets of the fairy-tale film-set castle. For one of the most-visited tourist attractions it was strangely deserted but as usual we had arrived in the lunch hour. It was almost too perfect but would be hard to remove from a tour itinerary.

We enjoyed our visit to Carcassonne and were looking forward to our move the next day to join Peter, Jenny and Ben at Camping La Figurotta, Bizanet.

We had learned to make use of facilities at the existing campsite before moving off to the unknown. On the morning of departure I searched in vain for the motorhome dump station at Les Peupliers but found none despite the *'Caravan Club Guide's'* assurances. At the far end of the field it did have four smart and separate bathrooms with toilet, shower and washbasin but they seemed inadequate for the number of campers that might descend during the summer season. The site manager made signals indicating I should tip it down one of the toilets. I couldn't be bothered and instead I filled up with water sneaking in just ahead of a German couple on the way to Spain in their motorhome.

We prepared Boris and parked him outside the campground on the road.

The campground owner was now observing all our proceedings as we unhooked the electric and put the slide in (after having removed gallons of water and a leaf mountain from the awning). I climbed aboard ready to move. The routine had become familiar to us.

Whoa! What was that red light on the dash? – The jacks were still down. I was so embarrassed that I confessed to Gail. The rest of the hook up went smoothly and we were soon wishing *'au revoir'* and on our way. Rather than trying to squeeze through the narrow village roads we retraced our steps onto the D11 then towards the A9 before taking the slip road onto the N9 heading southwest.

We negotiated the Narbonne ring road onto the N113, then after discussion the D613 and after a short while we turned right onto a narrow country road that wound its way round sharp bends up the hillside then over the top.

A red sports car, half over on the wrong side, came hurtling over the brow of the hill. It was so unexpected; my heart was pounding, and for the first time in Katie experienced a scary moment.

Soon we were on the campsite and Jan (the live together co-owner with Willem) came to greet us and said ritual chips would be served at 6.00 p.m. The campsite had a small number of pitches on a thin strip of land in amongst pine trees overlooking a forested valley.

As we unhitched Boris from Katie, Peter and Jenny made us some welcome coffee, which we enjoyed in their caravan with Ben the dog who had apparently been sick in the night. It was so good to be making friends and doing things with others.

Once settled, Gail and I decided to go to Narbonne and agreed to meet up again with Peter and Jenny for the chips and wine in the evening.

Narbonne, originally a Roman port and capital of their first colony in Gaul, offered much to look at. We particularly enjoyed the somewhat incomplete cathedral and surrounding buildings alongside the Canal de la Robine. They gave a real feel for their time (14C). We wandered the back streets searching for the Roman remains, and having decided not to pay the entrance fee to the underground museum found them on a street corner with the kids using them as goalposts.

Our most memorable sight in Narbonne was a large green sign *'Toutounet'* with wonderful diagrammatic pictures to explain the art of poop scooping: Take a bag.

Seize the Dejection. Turn over the bag. Throw the bag into the basket. It couldn't have been clearer – but the effect had been zero – there was poop all around.

In the evening we carried our saucepans to Jan and Willem's little hut to place our food orders off their menu and ate with Peter and Jenny in their caravan extending the enjoyable evening with much wine drinking (for those that did).

What a storm! Thunder, lightning and rain continued through Saturday night. I jumped out of bed to lower the satellite and TV aerial and we didn't sleep well. The next morning we had no electric as the storm had tripped out the supply. As the rain was lashing down and the normally stony, rocky campground was developing lakes, I decided against going out to switch on the inverter and started the generator – after all we did have to dry Gail's hair before she faced the world. It fired up beautifully and we had all the power we needed.

Our satellite couldn't get ITV1 for the World Cup rugby and we assumed this must be because we were too far south so we were pleased Peter could get the French TV channel on his Universal TV. England 24: France 7. We celebrated with Buck's Fizz and Danish pastries brought from Narbonne.

Peter and Jenny came for coffee and to look around Katie and were kind enough to say the furnishings and spaciousness impressed them. Peter tried the driving seat in a professional capacity being an ex traffic cop.

After lunch the rain ceased and we returned to reality. Gail vacuumed and dusted; I dumped the waste into the dolly then set off for the dump toilet, the usual arrangement of a large outside toilet. I cursed the gravel that was particularly deep and made pulling the heavy dolly hard work. On each of the five trips I left tracks behind me as I struggled up the incline that now felt like a mountain and arrived more and more exhausted at the *'WC Chimique'* where I prayed for an uneventful emptying.

Somehow the job was completed but as muscle fatigue developed I wondered how everybody else had managed before me, was I already standing on fertilised ground?

I wanted to flush the black tank on Katie but the campground only had push button kind of taps that require continual pressure and cause permanent indentations in the palms of your hands that might pass for crucifixion scars.

Then I saw an authoritarian notice that you must only obtain water by using the water containers supplied by the campsite – a green watering can with an extended nozzle – the same watering can that had been out all night in the storm. I filled our own water bottle then tipped it down Katie's toilet to flush the black tank.

Later Peter and Jenny came round for wine and around 11.00 p.m. we started watching *'My Cousin Vinny'* on the TV.

During the adverts Gail put the electric kettle on for coffee – and bang – the power went off. Peter and I snuck out and by torchlight hooked up to an electric post on another pitch, fully expecting to be found out by either Willem or Jan the next day.

The next morning, and halfway through showering, the electrics blew again. I checked with Peter next door but he had power. I tried two more sockets on the electric post but neither was active or we were blowing them as soon as we plugged in. The fuses were only accessible with a key. We were in trouble, what could we do not to incur the wroth of the camp commanders?

Gail was volunteered and with wet hair went to ask Willem to reset the fuses. She is my exocet of charm and I have never known her to fail and as a result he was also utterly charming about it. A 4-amps supply was really limiting.

Next my computer went down. I spent a couple of hours trying to coax it back to life, only partially succeeding.

Gail did some washing and met fellow Brit Emma at the communal sinks. She had been touring France and Spain in a large caravan and told Gail of the difficulties they had in Spain with 'tyre stabbers.' Whilst at traffic lights someone with a knife attached to his shoes had punctured their caravan tyre. A little way down the road they had to pull over and lo and behold a couple of friendly guys promptly arrived on the scene to help – yes one to help and the other to help himself to the valuables.

On Tuesday Peter and Jenny packed up to head for Spain (Gerona). I reluctantly relayed Emma's experience to them but they were undaunted, this was their grand tour, Spain was on the list, Portugal would follow. Should we go in tandem with them? We hoped we would meet again as new friends had become close friends all so quickly.

We decided on a tour of Cathar castles, those romantic and ruined medieval fortresses that range in a broad arc south of Carcassonne.

We set off via the eleventh-century Cistercian abbey at Fontfroide just south of where we were camped after another fifteen minutes going round the back streets of Bizanet trying to find the unnumbered left turn to Fontfroide. We crossed the D613 to find the car park of the abbey. We were told in the office that only the surrounding grounds (not the internal gardens) were open. We climbed onto the cypress-clad hillside and looked down onto the beautifully located twelfth-century abbey.

Disappointed not to have gained entry but anxious to get on to the Cathar castles, we left and headed back onto the D613 towards Thezan then left onto the D611 travelling south through the Corbieres wine country towards the magnificent Pyrenees. With the foliage on the vines turning gold and orange in the autumn sun the scene was exquisite with little traffic to spoil it.

Just before Tuchan we espied our first Cathar castle – Chateau d'Aguilar and turned left along the track between the vineyards above which it rose. From the deserted hilltop car park we climbed the mountain path to the ruined castle. For that moment alone our whole trip seemed worthwhile. No one was around, the weather was perfect with blue skies and a few white clouds, the views spectacular with golden vineyards leading to the snow-capped Pyrenees and Mount Canigou, and we were free to enjoy our imaginations as to what had occurred over the centuries in that castle.

We drove back through Tuchan onto the narrow mountain road (D14) towards Padern and *'The Route des Cathars'* and arrived alongside the mountain valley village of Cucugnan surrounded by more russet-coloured vineyards. This pretty village is dominated by a large windmill but is apparently more famous for its statue of a pregnant Virgin Mary and the Curé de Cucugnan, hero of Alphonse Daudet's book *Lettres de Mon Moulin*. Until the seventeenth century it stood on the French-Spanish border. We thought this an idyllic scene. Looking south we could see the mountain Chateau de Queribus and west Chateau de Peyrepertuse.

We went to Queribus first, just off the D123 (D19 from Maury) standing spectacularly isolated on a pillar of rock above a sheer cliff. What a defensive masterpiece – it was the last stronghold of Cathar resistance, holding out until 1255, eleven years after the fall of the castle at Montségur. Never reduced by siege, its role as a sanctuary for the Cathars ended with the capture of the luckless Chabert, whoever he was.

We looked northwest across at the mountains where we thought Peyrepertuse should be but couldn't be sure, was it merely a mountain ridge?

We headed back to Cucugnan then turned left along the D14 but even as we drove up the narrow winding approach road to Peyrepertuse, all we could make out was a tower on the top. After purchasing tickets from the little bureau in the car park we ascended the narrow mountain path up around to the back of the castle. Disconcertingly it dipped down at first, and we questioned whether we were really going to get up to the fortifications that appeared to be up in the sky. The guidebook said visits in strong winds could be dangerous and were forbidden during storms, as the ridge was a lightning magnet. The ever-narrowing path wandered through the trees on the side of the mountain away from the sun, and it cooled in the shade, making me wish I had brought my fleece.

It became steeper but eventually delivered us to a gate in the northern walls where workmen were making restorations, so we ignored this area and turned to the south, scrambled over the rocks past the polygonal building (for archers) and swallow hole and climbed to the western end of the rock and the fortifications on top (San Jordi). Passing through the buildings and an arch at one end, we arrived on a balcony with the most amazing view across the castle, the valley, Cucugnan, Queribus, the Pyrenees and all the way to the sea at Perpignan.

I was so enthralled I told four French guys they must go through the arch – *'c'est beau, le meilleur!'*

This wonderful castle had been occupied since Roman times from the start of the first century BC with a first mention in 1070 when owned by the Catalan counts of Besalù. It was remarkably self-sufficient for its location perched on the limestone ridge 800 metres above the vineyards and even had four sources of water. It had seen war and ravage, been in Spanish and French hands, marked the border for long periods up to 1659 with a token force of 'Morte paye' (local villagers), and sold to the state in 1820. What a remarkable history was contained within and about those walls; we felt quite inspired by the visit.

On the way down a couple were struggling up the path, the lady with a walking stick. I noted the English yacht club logo on his T-shirt.

'Not far now,' I encouraged, knowing full well the path became twice as steep.

We took the easy route back to Maury then onto the D117 to Perpignan and N9 northwards. We ran alongside the sea and the *Etangs* (what are they, lakes, lagoons?) to Narbonne and the campsite.

Next morning with Peter and Jenny having left, we considered whether to move on or do more sightseeing. With the sun shining again we decided to stay and moved the debate on to sea or mountains and decided on mountains. We had not explored the area north of Carcassonne.

To get us going quickly, we took the main road to Carcassonne (N113) then through Trebes, the D101 and Conques-s-Orbiel and the beautiful Orbiel valley, where deep ravines had scrub lower down and forests of chestnuts and pine higher up, to the village of Lastours. Ahead of us four castles appeared on the mountaintop, but how to get there?

We took a chance and followed signs to *'Villaniere'* and *'Belvedere'*. The road was extremely narrow and steep and squirmed its way up through the village, out the other side and higher and higher. We wondered whether the *Belvedere* would be miles away when suddenly a *'Camping'* sign appeared on the right. We parked outside the campsite and walked through some orchards believing there must be something at the end of the path. We came to a clearing and were surprised to find benches were arranged in a semi-circle looking across the valley, giving a magnificent view of the four Chateau de Lastours. We sat quietly, contemplated the wonderful scene and read from the guidebook.

The eleventh and twelfth-century castles perched on a ridge that plunged to the river on both sides were toy-like at that distance. To us they looked truly invincible and it was hard to believe they wouldn't have survived an attack. However, Simon de Montfort apparently did the deed in respect of the two originals (Cabaret and Surdespne) in 1211 and the other two (Tour Regine and Quertinheux) succumbed later.

From Lastours we headed out through the village and on to Roquefere, a fantastic site nestling in the valley but we couldn't find anywhere to park and headed left for Cupservies. The road climbed steeply away from Roquefere with a beautiful view back down to the village but the road was tortuous with no possibility of parking. The road presented some increasing challenges as it climbed steeply through the forest on a leaf-strewn road with a sheer drop to the right. The deep piles of leaves made it appear even narrower than it was.

We stopped momentarily to peer through the trees at the one or two isolated villages; their stone buildings clinging to the hillside, smoke issuing from their chimneys. A car, and the only one we had seen so far, went past working hard up the hill and probably amazed to see a vehicle parked.

The road narrowed, and climbed, the drops became steeper, the forest denser and the leaves thicker. We were on the edge of a ravine. Eventually we pulled into Cupservies, which comprised about four blackened stone dwellings and quite by chance over a wall we saw what we had come for, an 80-metre waterfall from the Rieutort stream that disappointingly only had a trickle of water. The area is apparently extremely poor and until recently, its people lived off beans, chestnut flour and the meat from their pigs working from ancient times in the region's metal mines. As we left the village still climbing, the road improved and we managed to find a place to stop and took photos looking back.

The road home was an easier route via the forest road D1008 then south on the D118 to Carcassonne where we ate a mid-afternoon lunch across the road from the station and great value. We arrived home as the sun set.

That night we decided that the next day we would go south.

Saint-Jean-Pla-de-Corts

I paid Willem the 53.75 euros for five nights. It was good value for a neat site in a delightful situation and marvellous surrounding countryside but tempered by the impracticality of low power (4-amp), push only taps, ban on hoses, lack of motorhome dump and having to wheel the dolly over deep gravel to the waste point.

We started to pack up and I decided to check the tyre pressures, which had hardly varied since we set off. Alarmingly the nearside, inner rear was down about 20-lbs. I connected our TRUCK AIR compressor to Boris's cigarette lighter for power and all went well for a while then it cut out. I couldn't find out why. All fuses seemed intact. By now a small crowd had gathered, always embarrassing when things weren't going right so I decided to stop and Willem suggested going to the petrol station on the main road.

As we hooked Boris up to Katie, Willem and Jan from the site explained it in step-by-step fashion to the small crowd and in particular to a Belgian couple who leaned in, lay down and stepped over everything to make sure they had the complete picture and hadn't missed any details.

Whilst I made a final departure check, Jan explained to Gail how we should take care in Spain to avoid trouble. We climbed aboard Katie and drove out waving goodbye to the small group that had assembled to see us off.

As we approached the petrol station we took a lane I had previously seen trucks using to avoid the canopy. I couldn't see where the air (*gonflage*) was and a truck had pulled in behind us. I decided to pull off the pumps to reconnoitre and then asked in the office and was told to look for *'the yellow.'* We did an about-turn and found a yellow *gonflage* but after we had worked out the bar to psi conversion we realised the pump couldn't produce enough pressure for our tyres.

With nothing accomplished we drove onto the A9 motorway heading towards Perpignan. Gail worked out which *aire de service* would have petrol and hopefully an air pump sufficient for truck tyres.

The motorway drive was steady and we eventually pulled in to Aire Lapalme-Ouest and cautiously approached the pumps. I had started filling with unleaded when the cashier came running out of the office waving his arms about. He was thinking I should be putting diesel in and was amazed when I told him *'essence.'* Gail had gone to look at the air pump to see whether it could cope but returned to tell me that it was no different so I suggested she ask the helpful man from the office. He came out again and explained we would have to go where the trucks fill with diesel. I could only do this by making a U-turn, to face the truck traffic coming directly off the motorway but fortunately it was light. We got the tyres back up to the correct pressure.

We took a break in the truck park then set off again.

We left at junction 43 on the D115 towards Ceret, drove through the village of St-Jean-Pla-de-Corts, turned left through a small housing estate then plunged down a narrow lane slightly wider than Katie and were about to round a left-hand, ninety-degree bend when a car appeared in the opposite direction and stopped quickly. He reversed to the wrong side of the road and we squeezed past. We then faced a concrete section fording a river with a sharp climb out on the other side. I wondered whether we were we going to ground the rear end. Fortunately we didn't and were soon up the hill on the other side turning sharply right and parked outside the office of *'Camping les Casteillets.'*

We received a warm welcome and found a pitch at the far end of a wooded and spacious campground with a beautiful uninterrupted view of Mount Canigou through the front window.

Setting up was no problem but as Gail hit the switch to send the slide out and I watched from outside she yelled: 'Come in QUICKLY!'

Inside Katie, on top of the slide, was an enormous cicada. Gail batted it to the floor with a cloth, and I despatched it outside, a four and a half inch whopper. It had flown in through an open window.

That wasn't the only wildlife on site – a piercing laugh alerted us to a donkey in the field opposite.

We started touring straight away and set off in Boris down the D115 towards Le Boulou and onto Collioure on the coast. A few days before we had seen it on British TV *'A*

Place in the Sun' and weren't disappointed. Descending from the surrounding vine- and olive-covered hills we found an idyllic scene of fishing boats bobbing in a peaceful harbour, a palm-tree-lined beach with sea-front cafés flanked by an ancient lighthouse doubling as a church at one end, a thirteenth-century Knights Templar castle at the other and a small town of narrow streets, small houses, shops and galleries.

We had coffee on Boulevard du Boramar, next to the beach and the focal point of Eglise-Notre-Dame Des-Anges (Church of Our Lady of the Angels), and relaxed but as we came to pay I had one of those heart-stopping moments when I realised my wallet wasn't in my bum-bag. What had happened to it? We scraped enough small change together to pay the bill. We were tense and raced back to the campsite hardly saying a word but thinking of the consequences the loss would cause.

A mad search revealed the wallet in Katie's door pocket. That day we had changed how we stored our valuables, what a great system – a raucous laughing and braying broke out from the dozen donkeys assembled in the field.

The next day was overcast so we decided to visit Perpignan. It was about a 30-minute drive using the N9 and the approach would be worthy of any American city with all the out-of-town shopping and advertising boards. As usual we headed for *'Centre Ville'* and then the nearest parking that happened to be Boulevard Wilson and under the park close to The Palais des Congres. Car parks we found were clean and smart (did not smell of toilets) with painted floors and clear markings even if the entrance and exits were usually tight. We emerged and walked in the wrong direction before realising we were lost. We retraced our steps and spotted The Palais des Congres with a number of delegates gathered outside all with their yellow nametags – was it a medical meeting like I used to attend in a previous life? – Those were the days.

Inside The Palais is the Tourist Office and we were soon in possession of a free *'Plan de Perpignan.'* We shot down some side streets towards the Cathedral St-Jean-Baptiste but it was all a bit boring – France has so many medieval towns; had we started to tire of them? We found a little cobbled square (Place de la Loge) near the Loge de Mer (Town Hall) that was once The Bourse and head of Maritime Trade (Gail read from the guide) and had a medieval sailing ship at one corner (we both looked up to see a diminutive sailing ship about the same size as a child's toy yacht). The building was started in 1397, The Town Hall during the thirteenth, sixteenth, and seventeenth centuries. Later as we passed we noted the *Maire* had equipped himself with some erotic mermaid statues in the courtyard in case he wanted something pleasing to look at though his window. Whilst reading all this we had enjoyed our coffees sitting outside watching the rest of the world go by and half-listening to two English women at the nearby table discussing everything from *Coronation Street* and *Eastenders* to the boyfriend situation. As the menu looked yummy and good value we wondered whether to stay and have lunch, but instead we decided to visit the main attraction The Palace of the Kings of Majorca at the top end of town.

We gained entrance for 4 euros each and were on our own climbing the stairs and slope up to the internal gate. The palace was more impressive from the inside than the outside and if properly restored would have had the feel of the Alhambra in Granada. We visited the empty rooms then climbed the tower. The panorama would have been magnificent but once again the clouds obscured the supposed view of Spain, the snow on Mount Canigou, and the sea, although we could just make out Perpignan and the lower hills of The Pyrenees. We returned to town and Place Arago for a look at the river.

It was then time to decide about lunch. In our usual way nothing appealed immediately and we wandered past le Castillet (City gate and then prison) and closer to the brasserie where we had coffee earlier. As we arrived and viewed the menu again I saw the same two English women had managed to move to the tables laid up for lunch. Whilst reviewing the menu I spotted another restaurant down a dark and narrow alley. It had a menu board at the entrance to the passage and a two-course lunch menu for 9 euros immediately attracted us.

The restaurant, (Spanish and presumably Catalan), was charming in a way that you know you're going to enjoy it. Initially our approach in French seemed to cause some confusion whether they were open or, wanted to serve us. After debate they appeared willing to allow us to stay under certain conditions but I just said yes to everything and sat down. We were served with a fantastic meal starting with country bread and aeoli then a warm mushroom salad followed by salmon ballotine stuffed with seafood mousseline, saffron and mussel sauce, timbale of rice with roasted red peppers and tomato. All assisted by a fine red wine for Gail followed by coffee. The walls were adorned with many photos of the chef with celebrities including bullfighters. This had been one of our most enjoyable meals and what value. As we left we had another look at the menu board and realised that this was something special as normally their dinner and a la carte were more like 40+ euros and the name of this wonderful restaurant – Casa Sansa. A pleasurable day and how warm it had been.

On Saturday November 22 I became Mr Grumpy because the final of The Rugby World Cup with England versus Australia was scheduled for the afternoon but I couldn't get ITV1 so I would have to wait for ITV world news at 6.30 p.m.

Gail suggested we go to Ceret, a few miles west along the Tech river valley, as it was market day and it would take my mind off the rugby. As we entered Ceret, and being Mr Grumpy, I ignored all signs for parking and headed for *'Centre Ville'* ending up in a street barriered off where the market began surrounded by people. I escaped down a side street where true to form we found a spot to park.

Ceret is a smart town and the bustling market spreads over a number of streets and squares lined by overhanging plane trees. We made some small purchases for lunch then enjoyed the usual coffees in a square near one of the gates in the medieval walls. We were joined by an oompah band in the full leather shorts and associated regalia.

The streets became remarkably empty as lunchtime arrived, yet cheering and shouting could be heard – the local bars were packed with TV Rugby watchers because the match had already started, what a dilemma – don't let me see the score.

Back at Katie, we had lunch then filled time with domestics – laundry, emptying tanks with the dolly. I then found we were parked too far from a working water tap and we had to use our modest 10-litre plastic container to fill Katie's 280-litre water tank and the taps were push button.

Gail then vacuumed Boris out, plugged into the cigarette lighter socket we thought defunct after the TRUCKAIR situation at Bizanet. Encouraged by this I tried the

TRUCKAIR and it worked. The TRUCKAIR must have a thermal cut-out.

Finally it was time for Rugby and I hadn't heard the score. What a match we (and unusually Gail) jumped all over the place, England were World Champions.

The donkeys proved to be noisy beasts at night, particularly if the male became detached from the females and when they escaped the field and surrounded Katie. No one seemed to be looking after them.

On another tour day we headed north up the *autoroute* to Perpignan then sharply west on the N116 towards Prades. The views were spectacular as we drove west with snow-covered peaks glistening in the sunshine.

In Prades we parked in the main square and were attracted to a modern patisserie with two tables beautifully laid with pink tablecloths. It seemed to be the source of all activity in the square. A queue of people developed to buy their bread but also to get the gossip of the day. We waited patiently and were served with large cups of coffee but also '*tarte a la crème*' for Gail and apricot and raspberry tart for me. We sat and listened as the population continued to come and go. Everybody was greeted by a cheery '*bonjour, Madame or Monsieur*' and then a '*merci*' and '*bonne journée*' on departure. We commented how polite the French were and all this for a total of 5.30 euros.

We found the minor road D27 and were soon at the Abbey St-Michel-de-Cuxa set beautifully in the mountains. A group of students sat on the steps eating their baguettes – the abbey was closed for at least two hours (it was lunchtime after all, even for monks).

We went on to Vernet-les-Bains; the road was spectacular climbing up through the forest and mountainsides to eventually drop into the valley. We took photos outside the *Mairie's* office and noted that the memorial was dedicated to cooperation with *'Anglaise.'* The snow-capped mountains were reminiscent of Switzerland or Austria and we worked hard to get the contrast right in the photos so the snow could be seen. The digital camera proved a godsend, as each picture could be reviewed and retaken if necessary.

We attempted a little tour of the old town; everybody was at home eating lunch except for the dogs that thwarted our walk. At two points we gave up only to have the vicious looking dog we were so wary of trot by quite amiably on our return route.

We headed out of town onto a minor road to Casteil, then St-Martin and the Abbey St-Martin-du-Canigou. After a couple of turns round the village looking for the entrance, we found a deserted hut at the bottom and a sign that said it should take 30-45 minutes on foot. The track was concreted and in season 4X4s could be hired for the steep ascent. The hairpin bends were continuous around the mountain edge, weaving up through the forest with streams crashing down below then over one ridge and up the next mountain. Eventually we reached the abbey sitting spectacularly close to the peak of Mount Canigou and looked down the ridges. It opened in ten minutes but the tour was by guide only and in French. We took a walk along the ridge to a viewpoint, Boris only just discernable way below. As we sat, we watched two blue overalled workmen burning autumn leaves, the smoke rising to join the descending cloud. The weather had changed for the worse and we decided to forgo the tour and retraced our steps back down the mountain.

We returned via Villefranche-de-Conflent, saw the fortress at the bottom and then Fort Liberia above the town apparently built to prevent aerial bombardment. The one thousand steps to the top were unappealing after the climbing we had already done and we headed home.

The following day the sun shone, Mount Canigou came back into view and a dozen donkeys were rounded up and squashed into a regular Transit van as though it was some attempt at the *Guinness Book of Records*. We started to think about Spain and even opened some maps out. We wanted to avoid Barcelona because it seemed a focus for attacks on motorhomers. I phoned the campground in Gerona that Peter and Jenny had gone to as it would only be a short drive and give us more time to negotiate Barcelona on another day but the owner said it was closed for the winter.

France takes its Sunday closing seriously and as nothing will be open there's no need to drive anywhere and on that logic it closes most of its petrol stations. Those that are open operate a self-service system with payment by credit card, which would be handy if they accepted UK credit cards, which they don't. This led us to finding ourselves marooned with Boris and an empty tank in a Perpignan petrol station. Gail put on her woman-in-distress look and found the best looking French rugby playing hulk on the station, melted his heart, and persuaded him to use his card in return for a cash payment from us.

As one sits in Katie wondering about the world and its injustices, any movement outside attracts attention so we were fascinated by an elderly gentleman carrying a gardener's kneeling stool and foraging for something in the football field opposite us. We thought he was probably picking mushrooms and, judging by the dimensions of his bag, very successfully. I went over to chat and found him to be an Englishman who had been resident on our campsite for five years. He told us how warmly the local villagers of St-Jacques had welcomed him. We decided to walk to the hamlet about a mile from the campsite and found a medieval village with an unrestored chateau. The village was showing signs of new money with a number of houses with scaffolding but it still retained that peeling wall charm. It had a *Boulangerie*, a *Tabac*, and a *Boucherie*. On a whim we went into the *Tabac* to ask if they had batteries for Gail's personal alarm, as being an odd size we had been unable to get them when shopping in the supermarkets or other news shops. The safety aspect of our move to Spain was in my thoughts. Amazingly she had them but for 6 euros each – the village wasn't that remote!

Whilst browsing her postcards of St-Jean-Pla-de-Corts, we noticed that one had a bathing scene at an *Etang*. She gave us directions to the lake and we set off on the short walk but after a half mile we were lost, so Gail asked two village ladies where the lake might be and received a fulsome but unclear answer. We walked ahead and the ladies shouted instructions from behind.

The lake was great and provided a marvellous adventure playground for children. We might have missed it so easily. As we had travelled through France we had explored intensively but what other treasures might have been revealed just around another corner. That was what drove us on. Others might spend more time relaxing we simply wanted the experience.

Back at the campsite I paid the lady in reception the 86.80 euros for our 7-night stay and we made ready for the next day's (November 27) departure to Spain.

Sitges

It was a true November morning, dark, damp and chilly with water dripping off the trees and Katie was enveloped in the morning wetness. As we prepared for the move into Spain, I was uneasy and I sensed Gail was too. I was also conscious we were being watched. From one of the smallest caravans we had seen and one of the few remaining left on site, a curious couple took in turns to peer through the curtains to scrutinize our every movement; but only one at a time, as there wasn't enough room at the window. They must have ached to see how much space we had in Katie and I wondered how they could lie down in theirs.

Despite our efficiency getting ready, events transpired to slow our departure – were we meant to go? The campsite owner was towing new mobile homes onto the field and a huge transporter, unloading the new and removing the old, blocked the campsite exit. I chatted with the owner to check on progress whilst Gail went in to return the security tag and emerged with the drivers of the *'Convoi Exceptionale'* who had been having coffee and maybe a small *pastis* for the road. The campsite owner and his family gathered in the doorway to wave us off and we hit the road only to be brought to a crawl all the way to the *autoroute* by a yellow JCB.

Finally we joined the motorway and made the long climb up to the French border control point just beyond Le Perthus. We didn't know what to expect. The right lane was indicated for trucks, the left for cars and the centre for caravans and coaches. We took the centre lane, fearing a height restriction, but it was coned off and merged with the lane full of cars that I noticed had police swarming all over them. Some inner voice made me quickly swing into an escape lane and into the truck route where we sailed through with no police or customs inspection at all. It took us by surprise. I looked in the mirror; but saw no sign of pursuing police or customs officials.

We approached the broad expanse of the Spanish tollbooths checking for a green entry light on the gantry above and ready to collect the ticket. As I drove in I saw the ticket automatically dispensed ready for collection, but as I opened my window in anticipation of collecting it I was amazed to see it disappear back in again all in one smooth movement, out and then in. The barrier remained in the down position. I stared in disbelief at the dispenser, wondering what to do. There was no button to press for a ticket only two slots; one at truck level, one at car level, but no ticket. Gail, also in doubt, checked whether I hadn't done something stupid. The logical answer would be to reverse out and go into another lane but as a barely legal width vehicle, we were tightly sandwiched in-between concrete walls about three feet high and twenty-five feet long with railings on top and Katie and Boris don't reverse. I saw a red button to summon aid. I pressed and it operated like a phone and over the little loudspeaker I could hear a ringing tone but no one answered and it stopped. I tried again and again. Meanwhile, alerted by our flashers those caught behind us reorganised their entry into another booth. I tried the red button one more time without success. Then we spotted an official blue-lamped car on the far side. Thank goodness rescue at last, we thought. The car crossed the motorway and came round behind us, through a tollbooth and out the other side to park about one hundred yards down the road on the right in a yellow crosshatched area. Two high-black-booted, dark-blue-uniformed officers with fancy red bobbles on their hats got out, adjusted their shades, preened themselves and looked around, oblivious to our plight. I sounded the horn, flashed the lights and Gail even tried waving out the window. They were unreceptive to our problem. We would have to try and reverse.

We donned yellow safety jackets and I tried to open my driver's door. It wouldn't open, as we were no more than a few inches away from the guardrail. Why had we bought such a wide RV? Gail tried the other door. It opened a little but the steps came out and jammed against the concrete wall. We were marooned in Katie. I was contemplating exit through the bedroom emergency window when Gail managed to squeeze through the door and between Katie and the concrete. I followed through the door but the concrete was too much so had to climb the guardrail. If we had eaten a bigger breakfast we'd never have done it. We got to Boris, unhitched the brake cable and started the engine to enable the power steering. Trucks thundered past into the neighbouring booths. Gail sat in Boris and I squeezed back into Katie and with a continual eye on the rear-view camera, reversed slowly, oh so slowly. We did about fifteen feet in a perfect straight line then I could see Boris was starting to 'crab' and his front wheels were locked in a turned position, yet we were not out of the concrete tunnel. Gail confirmed over the walkie-talkie that she had been unable to do anything to stop it. We were stuck again.

I descended from Katie, squeezed through the door, climbed the guardrail and together we unhitched Boris and I reversed him into the onrushing trucks. I raced round to squeeze into Katie and reversed her, Gail hooking Boris back on again without his safety chain and electric cable. Gail jumped into Katie and I wondered whether to drive back into the same booth and see if we re-triggered the ticket dispenser but couldn't face having to extricate ourselves all over again so put Katie on full lock and managed to get her into the next booth. A ticket came out and I grabbed it; the barrier went up and we were through and parked beyond the officials. They had spent their time pulling in trucks going about their normal business. We re-hitched Boris properly and watched in amazement as vehicles drove in and out of the lane we had got stuck in.

'Keith are you sure you didn't drop the ticket?' Gail asked.

'No, I never got near it,' I responded.

'It's windy, maybe it blew out,' she suggested.

'Then the barrier would have gone up. Anyway how would you ever get the ticket back? I couldn't get out of the driver's door and what a shocking state of affairs that nobody answered the aid button we could have had a major problem.'

'You don't think we did?'

There were mountains but a lot of heavy industry, the weather was initially the same but as we drove farther south blue sky appeared beyond the cloud and the mountains looked scenic. No one attempted to rob us but with this in mind we steered to the west of Barcelona keeping on the N7 as long as possible then down the A2 towards the airport. We encountered major roadwork, then an accident and were forced into narrow lanes squeezed up against trucks on each side. We took a ramp off, only to be brought back on, then tried again and drove onto the C245 and then the C32.

A strong wind arose and my arms began to ache as I struggled to keep Katie under control. We drove through a series of tunnels cut through solid rock – extraordinary examples of modern engineering and the digger's art; and I realised it would be easier without sunglasses. We passed one turn to *'Sitges'* and took the next to *'Sitges Centro'*. From the centre of Sitges we followed signs to Villanova along the C31 and then saw the flags flying outside *'Camping El Garrofer'*, our destination.

Christina, the girl in the reception, was in a miserable mood and admitted she was only waiting for December 18 when they closed and she could clear off for the winter. She informed us many had left because of the wind and plenty of spaces were available. We tried to cheer her up but she won and we became depressed as well.

The campsite was heavily wooded with open-plan dusty earth pitches beneath the sky-high pine trees. Roads led in all directions. It had a restaurant, bar and shop and a number of large blue and yellow concrete shower and toilet blocks. A few people were walking around and children were riding bikes. In the camping areas barely demarcated by the roads we saw an order to the randomness of the many hundreds of caravans parked close together with their large blue-grey, but now dusty and faded front room awnings with encrustations of bird droppings on their roofs, fronted by green ground mats. A myriad of water-filled bottles lined the walls of the awnings to prevent the wind blowing them away and an accumulation of chairs, cooking stoves, bikes, satellite dishes and other detritus lay around. I wanted to dump Katie's tanks. I checked the campsite plan and identified the location and found no more than grid over a hole at a sandy crossroad. We had no means to rinse around yet a car wash was twenty yards away. I had to check twice with the workers that the dump was what we thought it was. Eventually we parked on *parcela 345*, a hard sandy pitch with many trees.

We had lunch and then drove to Sitges in Boris to discover what an agreeable place it is. Surprisingly for a resort we also discovered that few spoke English or French and as we knew no words of Spanish beyond '*Que sera, sera*', and that included not being able to count, we found our brains had solidified and resorted to some primeval and probably unnecessary sign language in order to have coffee. On examining our change we discovered that crossing the Pyrenees had diminished the cost of milk in a cup of coffee by 1.25 euros.

Friday November 28 was a bright day so we went to Barcelona by bus. The stop was immediately outside the campsite and we waited with an Argentinean girl who helpfully told us about the service and the tour bus we could use when we arrived in Barcelona. The bus that arrived on time was a motorway express that stopped once in Sitges and wonderfully the driver spoke English. It was a delightful change to have someone else drive. We got off at the Barcelona terminus carefully noting this was outside Barclays Bank. The Argentinean girl told us to remember it since no one would know the name of our wonderful '*Mon-Bus*'.

We found the main square at Plaza de Catalunya and enjoyed coffees, sat in the sunshine watching the world go by and tried to decide between the two competing tourist bus services. We chose the orange bus because it did one tour of the town rather than splitting it into two. After paying our 16 euros each to the young girl driver we ran up to the top of the open double-decker bus in true tourist style. I eventually worked out the audio and persuaded the earplugs to stay in my ears, and wondered what was different about my ears as no one else was having trouble.

The two-and-three-quarter-hour tour was excellent and we stayed on throughout rather than hopping on and off. I attempted to take photographs as we drove past Gaudí this or Gaudí that, fighting the swaying monster, Gail laughing her head off as I bounced from one seat to another nearly ending up in other people's laps and narrowly avoiding low-hanging trees. The dynamism and nature of the city was reflected in the blurring of the resultant photos. By the end of it we were windswept and fast approaching frostbite even though the sun had shone beautifully.

Desperate to eat and warm up, we scouted around the Plaza de Catalunya, walked down La Rambla with everybody else then dived off into a little side street and found an admirable restaurant. The waiter couldn't have been more helpful even though he only had a few words of English and we enjoyed a three-course meal with wine (for Gail) and coffee for less than 20 euros. We walked off the late lunch with a promenade before we returned in our '*Mon Bus*' that had decided it wasn't an express anymore and would return via various outposts including the airport.

Barcelona was a great day out, an excellent place to visit, spacious, full of unusual architecture, with a real hum to it.

The next day we visited Villanova I la Geltru west of the campsite and parked near the port area next to the tourist office. The main Rambla seemed to go on forever and was more like a concrete motorway. Down a side street we investigated a market that had a wide choice of clothing, bedding, fish, meat and vegetables with noticeably lower prices than in France.

After lunch we headed back to the campsite. The caravans and awnings that looked dusty and deserted when we arrived had blossomed like giant flowers opening up and parents, children and grandparents were swarming round like bees. Close by us I watched one caravan as the men set up plastic tables and a mixture of plastic and beach chairs for eighteen diners, ladies laid the tables with plastic covers and many plastic bottles of Coca Cola, Orangina, wine bottles and sundry dishes, Dad got the giant orange bottled butane-powered paella pan steaming, others amused the children and dog and everyone was talking loudly and all at the same time whilst car radios played. A giant party atmosphere prevailed; the Spanish had arrived for the weekend. As the numbers grew, the steaming paella pan was rested as introductory conversations continued then finally around 3:15 p.m. Mama lifted the lid and steam rose into the trees as she scooped out paella portions onto paper plates for the children already sat at the table, then it was everyone else's turn. The women crowded round the pan to check proportions were suitable for their menfolk and passed the paper plates to them. There was enough for everybody and everybody had their seating place. Dad finally sat at the head of the table in the larger blue and white beach chair and inspected the plate of paella set before him, got up, lifted the lid on the paella pan to scoop up some more onto his plate and the meal began with noisy conversations flowing from one end to the other of the table. After a while Mama returned to the pan with a green paper dish and managed to scrape up the last remnants of paella for the little brown and white terrier dog that jumped at her legs in anticipation but when he smelt it seemed less enamoured so Mama introduced a few treats from her pocket and he was soon wolfing them and the paella down. He finished in double-quick time and showed his appreciation by cocking his leg and peeing on the paella pan lid that was resting on the ground. Two-hours later the children had sped off on their bikes but the adults were still talking. As we sat in the 20-degree sunshine and watched in awe, Gail put the kettle on for a modest tea and blew the electrics for about the fifth time; fortunately the fuse box was accessible to us. Once the Spanish had finished their meal and cleared the tables, it was time for the promenade or *paseo*. Groups set off to see what was new around the site and soon we were the objects of continuing attention. Oblivious to our presence each party stood by Katie in loud animated discussion pointing out this or that aspect of her construction or how Boris was attached, all circling to get a better view or coming up unashamedly close to examine some aspect in detail as one of them professed superior knowledge and like a tour guide explained it to the others. When they had exhausted their deliberations they passed by us with a knowledgeable nod.

Spanish latecomers continued to arrive in their cars at neighbouring caravans into the evening using their headlights for light and shouting instructions to each other about what to unload and put where and even as we went to bed they carried on partying.

Moncofa

On Sunday November 30 we decided to abandon Sitges and the weekending Spaniards.

I phoned our next selected site, Camping Vinaros, to ask if they could accommodate Katie, and for directions. They could fit Katie in but the directions were hopeless – 'phone if you get lost.' I tried to point out that getting lost was not part of our schedule, as we couldn't turn round to rectify any mistakes. It made no impression.

I reconnoitred how to extract Katie from her site without hitting trees, water or electric supplies as well as the large number of Spanish cars that had arrived and were parked in a display of random abandonment around the pitches. As soon as I started Katie's engine a crowd gathered on the corner to watch; fortunately the routing decision worked, and with Gail on the walkie-talkie we were soon in a position to hook Boris on. We then proceeded slowly through the campground, managing to avoid all obstacles. As we approached the exit barriers I wasn't sure Katie would get through without taking some of the tiles from the office roof. Gail went in to pay the bill that came to 56.69 euros for the 3 nights and to suggest they lift the entry rather than the exit barrier.

The barrier was raised and we drove out of the camp. It was a beautiful morning and we had soon driven up the hill and taken the sharp right-hander towards the C332 *autovia*.

The sun shone and there was hardly any traffic – was Catholicism so powerful that everyone was in church? The situation continued as we joined the A7 and headed south along the coast. As cars passed, the passengers took a long look at Boris's hook-up then up at us as they drew alongside. Because of the stories of motorway bandits trying to get you to stop by indicating something was wrong with your car or caravan, I kept a wary eye on them.

We made excellent progress; telling ourselves that we must travel on a Sunday again. We made our first Spanish fuel stop. The garage was under reconstruction or, resurrection but we found a pump we could access although not one that I could work. The Spanish instructions had something to do with keying in whether you wanted an amount in litres or money. The attendant was already confused because he couldn't understand why a vehicle as big as Katie's didn't need diesel. I explained the *'gasolina sin ploma'* (unleaded) was what was needed. He went away and suddenly the pump sprang into life; he had thrown the switch in his little cabin. As a result he was 103 euros better off but we weren't disappointed with the 0.811 euros per litre compared to 1 euro in France.

Poring over the map, we saw Vinaros was only a short distance away and with time in our favour decided we should go farther and called Camping Monmar at Moncofa about halfway between Castello de la Plena and Valencia. We obtained the helpful response that they would find a special pitch for us and directions were as in the *Caravan Club* book.

From then on the area surrounding the motorway was heavily industrialised and we were glad we had decided to push on. Eventually we left the A7 by exit 49 and were soon approaching the old town of Moncofa that sounds more romantic than the reality of old concrete houses. We knew we had to go through the town and then follow signs to the Tourist Office to get to the beach area.

The streets became narrower, the number of cars parked both sides increased, we made sharp turns first right then left and were in the middle of the town at a junction with no signs. With extensive traffic behind, Gail called over to a young guy and with her non-existent Spanish asked for the 'Tourist Office' but it didn't work. He looked up at Gail in the cab nonplussed. She quickly showed him her guidebook with the name of the campsite. *'Ah si!'* Arms waved in all directions but the emphasis seemed to be towards the end of the street that looked as though it had been closed by road works. Anxiety descended. I feared this was the moment when we would get trapped and be unable to turn, a scene we had pictured but dreaded. What choice did we have when the street behind was blocked by cars? We set off in blind hope that something would turn up.

Once close to the barriers I found to our relief that I could squeeze Katie and Boris through a sharp right turn to emerge from the darkened narrow streets of the old town into more modern open areas. We crawled along looking for a boat in a grass triangle, leading the town's traffic in a procession behind us. We saw a colourful painted dinghy on a grassy plot but as I came to a stop to make the left turn across the road I realised I would be turning into a cul-de-sac; we are poor in cul-de-sacs. I drove on to find there was a second colourful boat on grass where we could turn left. The traffic behind was so relieved and raced past; the squeal of tyres announcing their frustrated indignation.

Four hundred metres down the road we pulled into the long drive of the aquatic centre then sharp right into the campground. We disembarked, children gathered round and eventually two men appeared from the office. One was reasonably dressed, the other I would, doing him no disrespect, put in the Romany category. Their faces were pictures as they looked up at us in the cab; our interpretation was 'No way, *José*!' We tried to explain that we had called ahead. They spoke no English but Romany did speak some French. Eventually we understood it was the non-Romany's brother who had taken our call but was now having a siesta.

I was escorted down the site by my new friends and shown the problem. Aluminium frames used to support sliding sunshades enclosed each pitch. It was a great idea for a site that was bereft of vegetation but the frames were lower than Katie. Expecting uniformity throughout, we searched for a pitch where the Spanish labourers might have been too lazy to cut the support lengths and weren't disappointed; we found one pitch where the frame was higher. The tarmac road was narrow and turning onto the pitch tight but we had no alternative and I agreed it was worth a go.

I walked back to collect Katie, exchanging pleasantries with two groups of Brits who were finding this the most exciting thing that had happened in a while in the mode of 'Rather you than me, mate,' or, 'Serves you right for buying such a bloody big thing.' We unhitched Boris, the crowd swelled, small boys jostled to peer into Katie's cab as I descended and Spaniards who were about to start their Sunday lunch broke off to watch.

With help from everybody, including Gail on the walkie-talkie and Romany manfully bending one of the few trees sideways, Katie reversed into place perfectly and we even had room to put the satellite up between the aluminium bars. Boris went on the pitch behind and the crowd melted away.

With the kettle on and the chairs out we enjoyed the sun, with temperatures well into the twenties.

The rectangular-shaped campsite was newly established with 170 pitches back-to-back, all level and covered in stone chippings. Each pitch had its own washing up sink and water supply. In the centre of the site were a large heart-shaped swimming and activity pool (closed) and three smart heated toilet and shower blocks. It had a restaurant and supermarket (both closed) and a clean laundrette. You couldn't fault it on facilities but it was soulless. I noted from the brochure that *'When reception closes, a guardian keeps and eyes all night long'* – that was a relief then.

We met the brother Rafael in the office and on his recommendation we decided to take the short walk into the town where a medieval festival was taking place. We made a half-mile diversion to check the sea was really where it should be and then completed the 1.5 miles into town (village, I really don't know which).

Our attention was immediately attracted by a queue of children eagerly waiting to ride on braying donkeys. The rest of the medieval tents were initially closed but as 5 p.m. approached they started to open and the local population spilled en-masse into the town square to be entertained by musicians, stallholders, donkey caravans *'Caravana! Caravana!'* ridden by gleeful children and led by an Arab in full regalia. Everybody was in fancy dress, the streets were lined with straw and colourful banners strung across them. We were entertained by wonderful falconry displays with many large birds of prey flying off to settle on the chimneys and rooftops of the surrounding buildings waiting to be called in by their handlers. As Gail watched the display a bearded falconer quietly stood behind her his blue-leather-gloved hand outstretched above her head. To the delight of the crowd, a huge and magnificent eagle swept in over Gail, it's long talons outstretched to grab the bait. Gail instinctively ducked, smoothed her hair back into place, then burst out laughing along with the crowd as she turned to eye up the magnificent creature. It had a white underside with fine blackish stripes, grey upper parts and an ash-grey-and-white zone on the wings and a fearsome curved beak. We then enjoyed a medieval barbecued shish kebab with a wish made over it after the hand bell had been rung (our wish was that we wouldn't suffer from food poisoning).

After a long day, sleep came easily to me along with dreams of huge birds stealing shish kebab from Gail's mouth.

It was December and Valencia beckoned. We set off in Boris through Moncofa, rehearsing how we would manage with Katie when we left. We followed a truck, smaller than Katie, and he became stuck at the road works – how would we manage? We drove in on the N340, a busy road with an endless convoy of trucks, and parked in the centre of Valencia in Pla. De Tetuan.

After stopping for the usual *café con leche* we toured the city with 2000 years of history behind it that included being taken from the Muslims by El Cid in 1094 but subsequent to his death 5 years later, it slid back into Muslim hands. The buildings were elaborate, we loved the ornate ironwork of the hanging lamps and the dark-green-leaved orange and lemon trees lining the streets. The Municipal Museum held the first map of the city drawn by Padre Tosca in 1704. It took him five years because he used a tape to measure every city street – and was probably in bed for the rest of his life with bad knees.

As we returned to Moncofa, we once again explored exits for Katie but this time ended up in the orange groves miles from anywhere and then in the next village. Our conclusion was that we might have to ignore the ban on trucks and sneak our inconspicuous combination of Katie and Boris through the town centre in front of the police station. It was either that or we might be stuck for the duration of the road works.

Towards the end of the next afternoon as I finished washing Katie, a girl approached. Madalena told me she was the receptionist and I thought I was in trouble for washing Katie on the pitch but nothing was said. To limit any damage we talked about Katie and I invited her to look inside. We sat outside in the sun discussing worthwhile places to visit and she suggested an area in Valencia where the houses were built on stilts like Venice, and the salt lakes beyond Valencia.

The electric tripped out in the night, as the fan heater and electric water heater must have been on together and exceeded the 6 amps available. At 6:15 a.m. Gail bravely disappeared outside into the darkness of a freezing cold morning to throw the fuse.

We decide to visit the places recommended by Madalena. We reached the port area in Valencia but couldn't locate the houses on stilts, so headed for El Palmar a village in the middle of the lakes and 13C rice beds of The Parque Natural de l'Albufera separated from the sea by a slim slip of land. The ground floor of virtually every property had been turned into a restaurant, not small bistro types, but mass catering double coach-load types. Yet, each had beautifully set out tables, all empty and waiting to deliver the lunchtime eel and paella experience. To add to the authenticity, small fishing boats bobbed in the narrow waterways on each side of the village their tackle spread out on the roadside to dry. A fisherman, having landed his wriggling catch, offered a ride in his small boat but we declined – after the first corner it was merely acres of rice fields. One or two houses particularly struck us because they were triangle-shaped and we had seen similar constructions in Madeira. We drove back to Valencia along the narrowest of single-track roads dissecting the rice fields that we now saw were teeming with wading birds no doubt seeking their own eel feast.

Satellite TV worked fine and we heard about the terrible floods in France with 10,000 people being displaced and wondered what our situation might have been had we stayed another week.

Resources were constantly on our minds. In simple terms, 6 amps and 230 volts means 1380 watts. The fire on half power was 1000 watts leaving some for the TV, computer and anything else we needed to run. That meant the fire had to be switched off if the electric kettle was used and we would have no heating. At such times Katie's commodious interior took a lot of heating so the fire had to be in the bedroom or lounge as it was incapable of heating both if it was cold outside. The easy option would have been to use gas for the heating, water heating and refrigerator but in contrast to France only government vehicles could refill with LPG in Spain. So we constantly considered options. The answer was to find a site that had at least a 10 amps electric supply so the fire could be run on 2 kilowatt or head south in search of warmer weather. It was time for a coffee – 'Can you switch the fire off?'

On Thursday December 4, after another cold night, we woke to grey skies and wearily shopped at the local supermarket in Moncofa. Our enthusiasm, if there had been any, for Moncofa out of season, had gone and we decided to make a major move in the hope of warmer climes. I phoned a campsite just south of Benidorm at Torres de Playa. They could take us without problem and had large pitches, 16 amp electric and motorhome waste – could it be our perfect campsite?

Villajoyosa

What a night; wind, rain, the slide-out awning flapped loudly; we didn't sleep. It was Saturday December 5 and we had planned to pull out but with the grey damp weather, large puddles along the roadway and on the pitches as well as our wretched state we couldn't make the decision.

Rafael's brother came over to Katie and we assumed his hand signals and few words were an offer of help in getting Katie off the pitch. I pointed to the heavy clouds and he indicated that it was no better north in Barcelona or, south in Alicante and said we should wait.

I suggested to Gail that we emptied the tanks; it would be helpful if we decided to stay and would make Katie lighter if we decided to go. The waste dump was only a toilet behind the washrooms so I had to use the dolly and you already know my thoughts on that. At least the road was smooth tarmac and the toilet dolly fairly zoomed along, before I yanked it up the curb through the washrooms and out the back. It took five trips and one near overflow situation.

I finished at 11:20 a.m. and we decided to go, the clouds had lifted a little, and what else were we going to do – the site was somewhat sterile and had few people on it.

For once the slide awning didn't hold gallons of water as the wind had blown it all off and replaced it with sand so we were soon ready to hook up Boris by the exit gate. Rafael's brother drove down to assist manoeuvring. I reversed Katie off our pitch, across the one behind, made a sharp swing to the left, avoiding lampposts and the aluminium frames to the camp roadway and facing in the right direction. We paid the 82.5 euros for 5 nights on *parcela 128* getting a generous 25% '*descuento*' for staying more than two nights (it was obviously difficult to get people to stay more than one). With an '*Adios*' to Rafael's brother and '*Muchas gracias*' for all the help, we set off for Moncofa old town.

Our outward passage through Moncofa had been a matter of continual concern to us because although we had driven through in Boris every day we couldn't see how we were going to manage it with Katie and Boris together. The continual road works diverted traffic via sharp turns down ever-narrow streets bordered by old buildings from which lamps and streets signs jutted out and spider's webs of telephone and electricity cables drooped.

A small truck led us into the maze then stopped abruptly in front of us, the driver alighting to make a delivery, oblivious to the exasperated look on my face and the line of blocked-in traffic behind. The usual ritual of passing the time of day with the customer, mutual admiration of the goods to be delivered, lowering of the tail-lift and wheeling the goods inside the customer's premises and a discussion of the local football teams chances ensued. It was ample time to drink the coffee prepared by Gail and raise adrenalin levels to bursting point. We eventually squeezed past parked cars, satellite dishes and various other devices jutting out from ledges and the previously mentioned hazards and wondered how long we would have to endure it when suddenly we sprang out onto a clear road like a rat out of a trap and relieved raced for the *autovia*. 'Goodbye Moncofa.'

We drove along the *autovia* in the middle of a truck convoy then getting tired of reading the advertising on the back of the one in front I accelerated Katie powerfully past. The mountains started to rise in front of us and the scene became much more appealing. We ate as we travelled. One minute the sun broke through and we turned the heater off and put our sunglasses on then ten minutes later it was reversed.

After 125 miles, and an uneventful journey, we turned off at junction 65a for Benidorm and could see the many high-rise hotel blocks standing sentinel-like ahead protecting passage to the sea before we took an immediate right onto the busy N332 for Alicante to pass in sequence the go-cart track, crematorium, hospital and casino.

At kilometre 140.5, just over the brow of a hill, and on a busy single carriageway, I tussled with my conscience about turning left across a single white line and pulled off into the forecourt of a warehouse to do a semi-legal U-turn. The approach road ran alongside new houses then open land with pine trees and down a hill towards a blue sea and beach behind which and on the left was the campground.

As we pulled up outside the barriers, a jovial middle-aged British camper came round to the cab window to greet me.

'Hello, come far?'

'Yes.'

'Thought so, engine's throwing off a lot of heat.'

The office was welcoming and after selecting a superb 150-square-metres pitch on a raised part at the back of the campsite, we were feeling much more positive. We manoeuvred between the trees on to the pitch and the Dutch lady from the caravan next door came over to say hello and introduce herself. We also noticed three other American RVs and a lot of British registration numbers. Like an amorous young couple we took a walk along the near-deserted sandy beach, the Mediterranean lapping gently at our feet and a few white clouds floating innocuously in the deep blue sky. The beach was flanked by cliffs at each end; directly behind it was the campground and to the right two others all with no hint of commercialisation. A range of hazy-blue mountains completed the backdrop.

We felt fully relaxed, even in a holiday mood, and took the decision stay for Christmas as it would take more than a week to drive back to the UK and then we would have been in the grey of winter when we had just found what we had set out for – winter warmth and exploration.

The next day as we breakfasted, the sun rose over the horizon into a clear blue sky and for that day and with two exceptions, the rest of our one-month stay in Villajoyosa the weather was sunny, wonderfully sit-outside warm, and often sunbathing-hot, requiring the purchase of sun cream and lotions. Around 5.30 p.m. the sun would dip below the cliffs to the west causing a marked drop in the temperature.

Life on the campsite was different to anything we had experienced whilst touring France and the journey down through Northern Spain. It was the regular winter home to many Brits, Dutch and Germans and the Dutch lady next door said she and her husband had been returning for twenty years and stayed for five months, their family joining them over Christmas. A community atmosphere was generated and a great deal of entertainment to be had in people and vehicle watching. Consistently good weather induced us outside almost all of the time sunbathing, reading, listening to language tapes or, having our meals. The ranks of pitches in front of us with many shading trees reached to the beach and between we could see the twinkling blue Mediterranean. Our row was at the back and next to last and raised an important premium carrying three feet of sandy soil overlooking the rest. As we were on the route to the bar, toilets, launderette, swimming pool, dump station and car wash up the hill behind us, it put us in a prime position to meet passing campers and gather news and intrigue. Indeed we could have done a serious study of their gastric and renal functions.

Since the weather was relatively unchanging it would only provoke comment along the lines of 'Glad I'm not at home' so other topics became more prevalent and important. In a spirit that would probably be engendered in a prison camp, the hard-done-by community railed against the management. The number one issue was the electricity charges. This came as a surprise to us as we had been revelling in the 16amp supply believing it was included in the camp fee and we had given it little thought. It came as something of a shock to discover the substantial cost already incurred. Almost all on site were using gas to cook, run fridges and provide evening and night heating in their caravans and motorhomes. Katie had propane in her fixed LPG tank but it probably couldn't be refilled in Spain so once expended we couldn't run the fridge whilst driving. I referred the matter to the camp utility committee aka Cameron and Malcolm who set about the task with the vigour of bored holidaymakers nosing into someone else's problem. They studied Katie's system and decided it would be possible to connect an external butane gas cylinder to our existing pipes provided it had a regulator; I could get the bits from *'Ardys Bricolage'*. I asked the advice of another camper, who had an older American RV, about gas cylinders. He used butane through a connection similar to the one recommended. I pointed out the black carbon deposited all round the furnace exhaust on the side of his RV. 'It washes off easily. I do it once a week,' he explained. I wondered what damage the butane was doing to his burner and was unconvinced.

After a couple of trips to *Ardys* (every bit as good as B&Q) I started to understand the different connections for propane and butane but in the end had to accept that I was not going to get any for propane and settled on a butane kit from the huge Carrefour supermarket less than ten minutes drive from us.

I tried to fit the pipe and regulator myself but the attachment on Katie was much thinner and wouldn't fit. I wandered over to Cameron and he and Malcolm came over with a gas bottle. Between us we decided a piece of the old pipe would fit inside the new one after hot water treatment. New jubilee clamps were fitted and we were in business. Gail went into Katie to light the gas hob to check the new supply.

'Cameron, have you turned on the gas on at the bottle?' she shouted.

It was hard to hear her from the remote position the three of us had adopted – purely to be good witnesses should anything untoward occur.

I bought my own wonderfully shiny aluminium bottle of butane from the office for 9 euros and from then on we had limitless cheap heat, water heating, fridge cooling, hob and oven cooking and without carbon deposits on the flue.

I helped wash Cameron's car to thank him for his help. Later I asked him and his wife Lorna about their caravan lifestyle as they spent their winters on this one site. People were always asking what they found to do all day. They agreed that activities somehow expanded to fill the day. This was true for Cameron as a journey to the toilet could take a couple of hours by the time he had talked to everybody en route. He then usually forgot what he had set out to do in the first place. Cameron was a gentleman who knew everyone and it was he who had first met us at the gate on arrival. He reminded Gail of my father, always cheerful and with a willingness to help others.

Water on site was obtained from taps along the boundary hedge and some distance from us. The screw tap outlet was larger than any of the impressive collection of connectors I had amassed and I thought this might be another scheme to discourage use of hoses. *'Ardys Bricolage'* came to the rescue again and I hurriedly paced out the distance from Katie to the tap to find it was 40 metres – exactly the length of our two hoses combined. It meant we could fill Katie in situ.

At the top of the steep hill behind us was the toilet and shower block and alongside a *W.C. Chimique* as well as a drain in the concrete floor at the centre of the car wash. Why was it designed this way? Owners of some European motorhomes tried to drive up causing skidding tyres and scary moments particularly with larger vehicles or less skilled drivers. Most didn't bother and simply used their cassettes. The morning cassette parade was all part of our watching programme as we sat out in our chairs. Etiquette dictated no more than a 'Good morning' to those on the way to discharge whereas once emptied a full conversation could ensue. Katie couldn't be driven up the hill, and wouldn't fit on the space at the top, and I did not relish pulling the dolly up and down it.

As I was polishing Katie, a great excuse to get out of domestic chores and an excuse to be interrupted by passers-by, a Brit stopped and chatted with me about Katie and RVs in general. He asked me how I disposed of the toilet waste so I pointed out the disposal point at the top of the hill. At this point I realised the open bucket he was carrying contained his toilet waste, which he intended disposing down the dump. I suggested he should use something like a toilet dolly and showed him ours. He wondered where he might get one. Like us, it was his first trip. He was in a fifth-wheel trailer. How did he think he was going to do the disposal?

With our own dump status critical I spoke to a Wolfgang, a German who also had an American RV parked along and below us. He told me unofficially he had used covered drains set into the road behind us.

We manoeuvred Katie through the trees and dumped down our hose into the manhole in the middle of the road then re-positioned Katie in a better position on the pitch to give us more sitting-out room, lowering of the main awning and excellent TV reception. Dumping down the drain became a weekly event. Libby, who had watched our comings and goings from across the road, said she had to have a cigarette to calm her nerves. Brits Libby and David were in a Hymer motorhome and had been on site the previous week before heading farther south but failing to find a better site had returned. Our Dutch neighbours also told us they were impressed by our calm approach and we realised we had been subject to campsite scrutiny.

A visit to Benidorm had never made it on to our top twenty places to visit before you die yet here we were only fifteen minutes drive away. We went on our second day and wondered what the fuss was all about, but then found we were mistakenly at the closer La Cala Finestrat. It had a lovely little cove, a sandy beach and was trying to emulate its neighbour by building hotels that speared the sky and wouldn't be out of place where we lived at Canary Wharf. We did a little tour and were amazed by the cheap prices in the cafés and restaurants.

In Benidorm itself we thought ourselves lucky to find a meter free and having fed our coins in we realised that no one else had. It was a Fiesta day. We explored the old town with its many restaurants and *tapas* bars and along the Levante beach. It was crowded and everybody was promenading in the heat of the day. It was smarter than we had expected but probably changed out of all recognition in the summer.

'Coffee and Cake' was 2.50 euros but you could have had 'Coffee and Brandy' for 1.50 euros. At a beachside café we stopped to admire the samba dancers, none looked under 70, but they really were enjoying themselves. Well-behaved, heavy-leathered bikers and their girls gathered at another with a live rock band. At the end of the promenade a beach exercise class led by an elderly but fit looking Chinese in his red shiny tunic was about to start for anyone who wanted to join and a good many did. It was all great fun and we thoroughly enjoyed the sunshine, the blue sea and sky, and the crowds. We had caught the holiday spirit.

We spent more time than before relaxing outside, reading, chatting, strolling the beach as well as going into La Cala Finestrat to use the readily available and cheap Internet.

We would often go to La Cala Finestrat around 2 p.m. (having learnt the hard way that nothing opens before 1 p.m.) and enjoy a marvellous four-course lunch with wine for 18.40 euros in Peter's restaurant not more than five yards from the beach within a beautiful protected cove. We wondered how it would cope with all the people when the amazing skyscraper was finished.

On one occasion we were entertained, a word I use in the loosest possible sense, by a sixty-year-old woman in full Spanish dress singing (miming) and dancing (twirling) to pre-recorded tapes played through an amplifier on her shopping trolley. How she had the cheek to collect her money I don't know. I noted she didn't go to the restaurant next door. Maybe they paid her to stay away.

The 'O Solo Mio' an Italian restaurant also overlooked the beach and the menu of the day would provide us with Parma ham and melon, swordfish for Gail and pepper steak for me with fruit salad for Gail and ice cream for me all for 8 euros each.

A passing young dark-haired male entertainer provided amusement balancing on an exercise ball whilst juggling balls and then knives. The children were fascinated, particularly one young girl whose parents were dining in our restaurant. He worked hard to attract our attention but the trouser falling and retrieving whilst balanced on the ball seemed to be the icebreaker. Despite then dropping a few swords, the juggler drew a round of applause and a full hat of coins before he set off to his next engagement with his backpack of balls and swords and trolley to carry the exercise ball.

If by now you have the impression that we had turned into inactive deckchair-bound Benidorm bores you do not know Gail. Within a week Katie had been thoroughly cleaned and polished, Gail doing the roof, Xmas cards had been bought in Benidorm, written in the café over croissants and *'mermelada'* and dispatched to the yellow post box. We were fully replete with all manner of foods from the Lidl store strategically built next to the Carrefour supermarket and populated largely by German, Dutch and English visitors who, like us, found that labels printed in four languages were a better insurance against making some ghastly purchasing mistake. Lured by the single intelligible word of *'Sensitive'* I had the experience of spraying shave foam under my arm instead of deodorant. The metal cages containing the Monday or Thursday specials of ladies' underwear, cardigans, shoes, televisions, home cinema systems, kitchen knives, lamps, satellite dishes, fishing rods, everything you wanted but didn't need, drew in the bargain-hunting crowds. The Spanish shoppers had trolleys loaded high with single items – tomato sauce, baguettes, bags of sugar or bottles of wine that presumably would be served up in their restaurant or hotel.

The super modern Carrefour supermarket was a huge cathedral of commercialism. I have not seen an equivalent in the UK (although my experience is largely limited to the Isle of Dogs Asda and Becton Tesco), and provided an opportunity to watch the attractive mini-skirted roller skate girls who dashed around and obtained a price if the cashier found she hadn't one, or exchanged things that were damaged.

On Wednesday December 17, we decided on a tour to Guadalest. We headed north along the coast on the N 332 and were amazed just how far the roads into Benidorm spread out. Before we knew it we were in Altea, hoping for a pretty little village but encountered a continuation of Benidorm. We turned inland to Callosa d'en Sarrià in the mountains but again were disappointed. As the road snaked up the mountain however, the views become more spectacular as the landscape changed and Guadalest emerged, perilously perched on a rocky outcrop.

Tour buses were parked in the car park and soon an attendant came over to collect our money, well that's what we thought he was. It was much colder so we enjoyed coffee and cake sitting by a roaring fire in a mock alpine hut along with a blue-rinse tour party. It made me think – why do women out-survive men? Is it the work strain the men have been through; in which case the ages should level up as more and more women work full-time?

We walked into the little town, proceeded through and explored the four floors of the Orduno House, the highest house, and then descended the antique stairs to the castle and church. We ended up in the square and one of the many cafés where we sat by the great stove and enjoyed a perfectly cooked meal. An old gentleman from the village wandered in, stood with his back to the stove as men do to get himself warm, before he went on his way untroubled by the need to purchase anything. Our return through Polop, La Nucia and Finestrat provided some more fine views. Guadalest is said to be fully functional with a school and visiting doctor but being close to Benidorm is a tourist magnet but it may also be the closest thing to the 'real Spain' that many tourists see.

I ran our generator for thirty minutes to keep it in trim whilst I chatted to Daniel who had been on the site for a year. He, his wife, and three lovely young children were camped in the row behind us in a 38-foot, fifth-wheel American camper that they pulled with an American SUV (mini-truck). He told me he was a Country and Western singer and worked the clubs in Benidorm. His brutish English terrier Alfie, stood with him and wagged his tail a lot so I tried to be friendly by offering the back of my hand for him to smell. The sight of teeth caused me to withdraw it quickly. The family would go back to England for a week so Daniel was planning to stow all the Christmas presents in the lockers, as all the children believed in Father Christmas.

Daniel and the family still only ate 'British' food and had not visited anywhere beyond Benidorm. The guys at work had been trying to persuade him to try Spanish food so one day he bought some '*jamon ibericó*' but was unsure whether it was cooked. We gathered round the packet and Gail assured him it could be eaten without cooking, but he was suspicious and his vegetarian wife was no help. Gail and I ate some and Daniel did a 'John Gummer' and fed some to his two youngest children. Then he tried some himself but spat it out. The '*jamon*' was destined for the dog bowl and Daniel's experiment with Spanish food was over.

Gary, Diane and their two children Jack and Emily, lived full-time in their thirty-five-foot Damon Challenger but moved on from place to place. The children aged around seven and ten were delightfully well-behaved. I asked Gary about their education. He was confident they were well-educated with all the experience of going to new places, meeting people of diverse cultures, educational projects they set for them, lessons from Mum and on the Internet, and their own little money-making industry selling CD recordings of TV soaps to motorhomers without satellite TV. They seemed well-adjusted and mature.

It was December 20 and the sun shone brightly. After morning chores and an outdoor lunch, we set off in T-shirts and shorts for a two-hour walk towards La Cala Finestrat and Benidorm along the cliff tops. We were amazed that such undeveloped and wild scenery could be so close to skyscraper buildings. There must have been rural life here at some time because we saw occasional man-made terraces, a small orchard, cultivated land and tumbledown stone cottages for the farmer or shepherd. The sea was sparkling blue with lighter then darker green patches and the white crests of the waves were hardly discernible. At one point we had to descend down to the large waterside rocks and boulders where a group of nude men lounged, and clothed men nonchalantly stared skywards as they cruised around.

'It's a gay beach.'

'They're just sunbathers,' said Gail.

I linked arms tightly with her so they could see I was of the other persuasion. We ascended the rocks on the other side and continued up the sandy path to the ruined stone 'Castillo' that sat dejectedly on the cliff top amongst the pines. The vista before us encompassed the burgeoning skyscraper pinnacles of La Cala and the cliffs at the end of its beach with smaller white apartment blocks clinging to the rocks. Farther round we could see the curves of Benidorm's two beaches and the Legoland forest of hotels behind the promenades. To the north was the whiteness of Calpe and then Peñon d'Ifac, a huge lump of limestone rock protruding out of the water like a giant bathing hippopotamus. Behind La Cala and Benidorm mountains rose abruptly and even though bathed in clear sunshine had a remote look to them.

A Spanish gentleman smartly dressed in his dark blue slacks, white shirt and tie, and polished black moccasin shoes, ascended with his dog from La Cala Finestrat. We greeted each other in Spanish (the language tapes were working) and agreed the view was 'very beautiful'. Conversation beyond this was stunted but he understood we had walked from Torres. As we descended the steep rock-strewn path we wondered how he had managed it in his best clothes.

After lunch in La Cala Finestrat, we returned the way we had come and recovered over tea chatting to our Dutch neighbours. She told us she was 74, but didn't look anything like it, had cancer seven years ago, was operated on in Benidorm with no waiting and swam every morning in the sea at 7.55 a.m. precisely. We met her daughter who had arrived recently with the grandchildren and would stay to the end of January.

A US motorhome with a Swiss registration – a Monaco La Palma identical to ours in all but the name – had pulled in close behind us while we were away. The next day we chatted to the young Swiss owners and found that they ran a driving school using the RV. Franz had deliberately removed the ladder at the rear of the RV to deter robbers who might climb on the roof and administer gas. On the way down they had stopped overnight at a service area and had been woken by someone breaking in the side door. Their lock had been damaged but to their relief entry was foiled.

Another seasoned camper had told us how he was 'gassed' and robbed on the way down when he stopped at a motorway service area. They broke into his Fiat Ducato based van without damaging the locks so he assumed they must have had keys. Others reported forced entries as they had slept, others, simultaneous multiple robberies, and a degree of indifference from attending La Guardia. These stories were told and re-told as campers passed from one pitch to another. Many were sceptical as to how anyone could feed gas into a motorhome at the right dose not to kill the occupants. Were the perpetrators moonlighting consultant anaesthetists? If the gas was volatile how did they avoid the naked flame lurking in the air intake behind the fridge? Maybe peeps were so tired from driving or having a few beers perhaps they simply slept heavily? But suppose they broke in then chloroformed the sleeping passengers?

Our ex-German military friend Wolfgang dealt with matters in a different way. He told us of the time in a new campervan and on their first trip into France he and his wife had parked up for the night alongside three caravans and were awoken by a huge bang and the camper rocking violently. Wolfgang drew his ex-army pistol and jumped out to find a car transporter had scraped his trailer the full length of Wolfgang's campervan, gouging his lockers. Wolfgang immediately reverted to military mode and shot out two of the truck's tyres so he couldn't drive off but then found the driver was slumped over the wheel fast asleep. A French speaker provided an accident report, filled it in then Wolfgang persuaded the transporter driver to sign it, presumably at gunpoint.

Others said they had a gas attack but found out it was down to the previous night's curry. We all began to feel relieved we had made it and felt safe where we were with our wagons circled defensively only breaking out occasionally for a full English breakfast, or 8.50 euros three-course 'Menu del Dia' with bread and wine included. The early capture of Saddam Hussein also brought some relief.

Although some had left the campsite, either to go home, or move on to other places such as Morocco, more arrived in the run-up to Christmas. Friends of the Swiss couple came with a van, motorcycles on a trailer and a caravan. All the vehicles went on the site behind us and the trailer was pushed onto the site next to us. We began to feel our space was being invaded. It was easy to escape and we couldn't believe that on December 22 we could walk the beach in scorching sun, sit outside for coffee, then around 5.30 p.m. walk to the rocks at the far end of the beach to watch the beautiful sunset knowing the UK was suffering snow or rain and was cold.

On December 23 we donned walking shoes, T-shirts and shorts and set off up the hill and over the cliff-tops to the south of us towards the village of Villajoyosa. Below us, surges of creamy blue water crashed onto the rocks. Out to sea were offshore fish farms and the occasional fishing boat bobbing along in that sea of blue. At the top of the hill, invisible from the campground, were partly finished yellow apartment blocks but after that a narrow track ran behind the older cliff-top houses down the rocks to the seashore then into the large modern port of Villajoyosa. Here, boiler-suited fishermen stretched out and repaired their nets. In another area men hammered blocks into place prior to bringing a ship out of the water and amongst some rotting hulks other boats were under repair or construction.

Towards the town the terraced houses on the seafront were all brightly painted; some blue, others yellow or green, supposedly so fisherman could identify their own. I couldn't think of a logical reason why a fisherman would wish to pinpoint his own house as opposed to the town itself when out at sea. What comforts could that bring him that his house was still standing or perhaps it was to check there weren't visitors his wife wasn't telling him about?

Along the seafront lunch menus cost around 9 euros; I noted one glum looking waiter outside his, bored through lack of custom. We stopped to chat to the English owner of a Spanish restaurant who advertised Christmas Day lunch for 30 euros including wine with a traditional menu. She had only four places left.

We climbed the hill into the old town where houses were built on the castle walls then into the church square and the surrounding streets. This was a different world of narrow streets and brightly coloured, three-storey Moorish houses. Occasionally a door had been left open and you could see the family sitting round in the dark cave-like room, a wood fire burning with an oil-cloth-covered table; the only light coming from the large TV that dominated the room. In the shadow of town we also felt the cold and decided against visiting the chocolate museum. We descended through some flower-bedecked ornamental gardens to the beach and had a splendid lunch of carrot salad, chicken risotto, and pear tart at *'Le Patio'* restaurant.

We walked back over the cliffs remarking what an idyllic spot our little campsite was and sat out in the sun until it disappeared over those same cliffs. The next day was Christmas Eve and the sun shone once again.

After coffee we decided on a drive to Calpe about 30 kilometres north along the N322; coastal urbanisation was going on everywhere. We parked in an unmade lot, behind a block of flats at the entrance to the Parc Natural of Peñyal d'Ifac and the impressive rocky outcrop we had seen on our cliff-top walk.

We started the ascent up a marked and fenced trail through the thick pinewoods – the views down to Calpe and beyond were already fantastic. Gail, not anticipating a rock-climbing expedition, was wearing her brand-new leather fashion boots. Her concern was not that she would slip but that she would scuff them.

We walked quite a way up and arrived at a slippery tunnel through the rocky outcrop. On the other side the path was described as 'dangerous'. We discussed with two descending Brits how difficult it became. They said the next section was the worst but was roped – we decided to go for it and made it across the polished, sloping rock. We climbed to the top of the first peak among the nesting sea birds but could then see people at the top of the second peak (1089 ft high) and decided to come back another day with proper shoes. We had done half of the supposed two-hour walk.

Back in Calpe we visited the port, the beach and market then went up the long hill to the old town and church. The streets were narrow and photogenic but everything was closed for lunch – even the restaurants in the old town. We descended to be saved by a café whose owner spoke to us in a British accent – it didn't feel the same.

Christmas Day started with brilliant warm sunshine. We started with a special breakfast in Katie then went to the beach in shorts and cut-offs like we were enacting some Australian travel brochure and phoned our respective parents to wish them festive greetings and ask about their weather.

We had coffee and chats with neighbours in the warm sun. More friends of our Swiss neighbours arrived – a caravan and two more bikes – and were parked on the site next to us. Our Dutch neighbour Lena came over to check that we were OK with the increased activity around us.

We had made no plans for lunch so decided to go to Benidorm. Shops, bars, cafés and restaurants were open. The promenade was packed with families in their Sunday best: Grandmas, Grandpas, mothers pushing babies in massive strollers like giant go carts, children experimenting with new toys, Dads ogling the topless Brits on the beach, all enjoying a walk in the sun.

In an old town *tapas* bar we forced our way through the throng to get two high stools and pulled ourselves up to the glass counter that was full of an amazing variety of *tapas*: – Serrano ham with mushrooms and garlic on top of toasted bread with olive oil, Serrano ham with blue cheese spread and anchovy, black pudding and pine nut, broschetta of gamba, mushroom, pimento and Serrano, salmon wrapped round tuna, bacon round dates, battered courgette cheese and Serrano ham, and many others we can't remember. We selected four to start, with drinks. The lovely girls behind the bar were working flat out, yet remained cheerful, each handling many customers' orders at the same time without apparently writing anything down. We thoroughly enjoyed the atmosphere and good-humoured Spanish crowd that surrounded us. We ordered more food and eventually threw our napkins onto the floor to join the paper mountain that was growing underneath us in an otherwise spotless bar.

Later, we sat outside Katie in the sun, had tea and chatted to the Dutch daughter and son-in-law of our neighbour Lena and later Cameron – a perfect day.

The days after Christmas we did lots of household jobs and I helped the young man in front fit a new exhaust to his VW camper – well I offered and sprayed my WD40 when called on to do so. More people drifted away, we began to feel like residents.

Almost to break the routine we drove the 20 miles to Alicante to be tourists for a day. We wandered the streets almost aimlessly without consulting the guidebook and ended up in the old quarter below the castle and in a square with lots of smartly uniformed policemen going in and out of an ornate building 'an eighteenth-century palace of golden stone and baroque façade flanked by two towers' and now the Town Hall. Next door the busy and homely restaurant's massive four-course 'Menu del Dia' for 9 euros, including bread, a bottle of wine and soft drinks, proved too big a temptation.

We felt more touristy after lunch and with renewed energy walked the long esplanade down to the beach then used the grey ironwork bridge to avoid frenzied Spanish drivers returning to work after their two-hour lunches, only to have to traverse a dangerous footpath alongside the six-lane highway to a gain entry to a dank tunnel. It was similar to the Greenwich tunnel at home and in similar fashion led to a lift but this one took us to the rock and Castillo de Santa Barbara.

We livened up, as the sixteenth-century castle was surprisingly impressive. We leapt into a turret with a hole in the bottom that you could look down (or was it an ancient toilet?) it seemed to be suspended away from the walls with little support. We felt queasy. The 360-degree panorama would have included both sea and mountains but as usual a dark cloud had descended to cheat us out of it.

At the bottom in the old quarter (Santa Cruz) we read about the eighteenth-century Iglesia de Santa Maria's baroque façade and its fourteenth-century Muslim origins but were more interested at looking back up at the castle and the turret where we had stood, reliving the stomach-churning experience. Then the sun burst forth and it seemed right to wander the endless palm-tree-lined marble esplanade. We unexpectedly liked Alicante.

'Get off the phone and please tell your mother not to call us on the mobile.'

'I can't do that – it's New Years Eve

'She says she is calling us so that we save money on our bills. She has checked and it's not costing much at all.'

'Of course it isn't – we're paying for all the bloody incoming calls.'

It was a problem that was difficult to resolve. Telling your mother-in-law not to call her precious daughter did not rank high in the endearment stakes.

Not knowing what our expenses would be, Gail kept a daily record as a spreadsheet on the computer and we seemed to be living within our means. Before leaving home we had made sure all bills would be Direct-Debited but as mail was being held for us there were some expenses we didn't know about until we had access to the Internet in La Cala Finestrat. Mobile phone bills had leapt six-fold.

We went down to the English-owned bookshop in La Cala Finestrat for a newspaper where we bought a Spanish language book and asked her about phoning home. She recommended a phone card that promised 200 minutes of international calls for only 6 euros. I was sceptical so immediately tried the card in the *'Telefonica'* box outside and a robotic voice confirmed the tremendous value.

'So, you can speak to your mother for two hundred minutes and it will only cost us six euros instead of one hundred and fifty to two hundred pounds on the mobile. It's hard to believe.'

I immediately called Nigel at East Coast Leisure.

'Happy New Year to you as well. Have you received those satnav disks yet?'

He hadn't but we still had 194 minutes left on the phone card.

'Who shall we call now?'

We rang a few friends and then my accountant to check my tax affairs had been dealt with on time. They had but did I appreciate that tax on the proceeds from the sale of my business had to be paid immediately? She would know the exact sum in a few days. I spent the next half hour moving money around from one account to another. The new phone card had proved a little miracle and we still had 130 minutes left.

We sat out in the hot sun and I swatted up on my Spanish, but faced constant interruptions from passers-by all curious to ask about Katie. By now we had realised that almost all of these conversations had one end point – a request to look inside. As our Dutch neighbour suggested at times, it was like being in a zoo. Brits were subtle and never directly asked hoping you would invite them in, the Dutch were polite, the Germans asked directly, the Spanish brought family and friends and stood on chairs so they could peer through the windows even when they knew we were inside.

Dani's bar at the top of the site provided the venue for the New Year's celebrations. It was packed. The campsite-owning family presided behind the bar and handed out free champagne and twelve grapes to eat as the clock chimed down to midnight. We gave and received lots of hugs and kisses, and as a result the 70-year-old part-time chef took a shine to Gail and bestowed her with an old faded photograph of him in his twenties. One hour later we did it all again to correspond to GMT. We also discovered our friend Cameron was an accomplished Karaoke singer.

Two days later, on January 2 and in a bid to get our bodies mobile again in the heat, we returned to Calpe and completed the walk, challenging in places, to the 1089ft summit of the Peñyon d'Ifac cliff the highest in the

Mediterranean, but this time we were prepared with proper walking shoes. The last part required a real scramble up the rocks to perch on the top of the sheer cliff face and for once we were rewarded with clear skies and fantastic views along the coast to the north and south, inland to the mountains and out to sea.

Back at the site, all the Swiss had left, leaving us surrounded by space as we were originally. We decided to make a move the next week on January 7.

Still without satnav discs, we went to Benidorm one evening to have a meal and buy a map of Southern Spain but found it to be crowded with little parking space available. We bought our map, had coffees by the sea in KM's club then did a promenade along with the rest of the world. In the old town white plastic chairs, some already occupied, were laid neatly along the pavement and at the edge of the roads. We wandered through the old town to the port and here saw many bands and floats assembling for a procession.

We went back and bagged ourselves a couple of the still empty chairs when an official looking lady came over and demanded 3 euros for the seat – no way – you can see just as easily from the pavement.

Eventually down the now crowded street, a band appeared at the head of the parade then a series of floats, more bands, an elephant, camels, llamas, beautiful horses and elaborate costumes. Dinky market garden tractors, usually orange in colour, pulled the float trailers. They were driven by Josés and Juans who didn't appear to have gone home after work to get changed and their tractors had come straight off the field never having had a clean in their long, long lives. Many of the floats were quite exquisite with large numbers of children and adults sitting in tiers with costume themes of Arabian nights, pirates, space, scary witches, fairy princesses, Andalusian traditional dress interspersed by dancers and bands. One float appeared in the distance to have a bouncy castle on it but as it drew closer it was possible to see this unintended effect was being produced by a punctured tyre that no one seemed too concerned about. The Three Wise Men were on camels, Roman centurions marched, belly dancers shimmied alongside prancing horses but the loudest cheer was for the street sweeper in his official uniform walking behind the elephant to collect any droppings.

Now, delightful as this all was, it was not the point of the parade – all the float people and those walking alongside were expected to throw boiled sweets to those lining the route, children especially. So the way to get the best cheer was to hurl more sweets. Sitting immediately in front of us were three Spanish ladies in their late fifties or sixties who were determined to have the biggest collection of boiled sweets by the end of the evening. As the float approached hands stretched out to those on the float imploring them to empty their straw baskets of sweets but there was a great deal of competition – hands were raised along the whole route – the success of a float was judged solely on the offloading of the sweets. As each shower of sweets was launched bodies moved in all directions to gather in this rain cloud of hard sugar candy. As the ladies leapt forward we behind grabbed those sweets flying over their heads and took particular delight retrieving those that fell into the chairs of the ladies who had now left them to seek bounty in the roadway.

Once the initial response was over, everybody in the front row seats waited for the float to pass then surged forward into the road to grab those sweets remaining uncollected. The children were particularly good at this phase, being a bit faster off the mark and carried with them large plastic bags to collect the sweets in. A little black boy in front wasn't quick enough to catch his share but a pretty little Spanish girl offered him some of hers and his smile brightened the whole street.

In amongst the sweet collecting, everyone was getting showered with confetti from those on the floats who had no sweets to disgorge. They were met with disparaging groans when it was realised their scatterings were devoid of sugar.

The sport developed throughout the evening becoming more aggressive but the mood was good and the ladies in front only retired when their coat pockets were full to overflowing. In what seemed no time at all our own pockets were stuffed to the brims and famished we headed for the Hong Kong restaurant where 11.95 euros bought two four-course meals, a bottle of wine and schnapps to finish. We drove home through La Cala Finestrat where at the crossroads the girls were out with a particularly impressive festive display of white boots and short skirts.

January 6 and our month's stay had come to an end. We prepared Katie and gave her a final polish before dumping the tanks then turned her round and parked on the treeless pitch behind us ready for departure.

We settled the 658 euros bill for 33 days and gained some schoolboy enjoyment because the final night's electric was 'free'.

We took our last beach walk as the sunset and wandered the campground saying goodbyes and handing out our sweets to the children.

Almerimar

'Look at the sun shimmering off the sea, that's what we came for.'

'It's not.'

'Well of course it's what we came for.'

'No I don't mean that. Your sea has trees in it.'

'No.'

'It's a sea of plastic!'

We were looking down on the land surrounding Almeria. To our right were dry bare mountains and on our left valleys sloping towards the sea. The God of horticulture had cling-wrapped the land between, only an odd tree, hillock or building had punctured the sun-reflecting polyethylene.

'It's ghastly.'

'It is, but I bet that's where a lot of the UK's tomatoes come from.'

We had set off early that morning from Villajoyosa via Alicante, Elche, Murcia, and Lorca and nowhere in particular. We stopped to buy a map at a service station that didn't sell maps and only had diesel. At the next one the petrol pump didn't work so Gail went in to seek assistance and spoke slowly to the attendant in a way that you do to someone who hasn't had the benefit of an English education to find that she had. She also thought our motorhome was beautiful and 'You could live in it with a dog.' I hoped she meant what she was saying and wasn't referring to the driver.

We passed Almeria and turned off towards Almerimar, where a new concrete road was fronted by a newly-built supermarket, shopping centre, hospital and agricultural produce distribution centres all, no doubt, the result of EU subsidies and tomato purchasers for Tescos, Waitrose, and Sainsburys.

We left the cliff-top plateau to descend a steep hill, the sea glistening below along with more plastic. The road at the bottom was newly built and stretched for miles along a soulless beachfront. We turned off down a bouncy dirt road and round a corner onto a four-lane dual carriageway with a blue-tiled central reservation of palm trees and fountains and at the end the white-painted reception building of Camping Mar Azul. Seeing a line of queuing motorhomes and after driving over 230 miles I was relieved I had phoned ahead to be assured they could accommodate Katie and Boris.

'*Buenas Días. Soy* Keith Mashiter. *Tengo una reservación.*'

'You have a motorhome then take a map, find a site and come and tell us, do not go on roads E or F.'

I staggered out with the site map to see crowds of people examining Katie and Boris.

'How did you get on, are we all set?' Gail asked.

'We have to find a site then come and tell them.'

'How big is it here, there seem to be a lot of people queuing?'

'Let me have a look at the map. Bloody hell! One thousand pitches.'

We were tired from the journey and exhausted after walking two circuits of the serried ranks of 1000 caravan and motorhome pitches looking for one that was vacant and might accommodate Katie and Boris. The only possibility was one on the prohibited row F but even that was too small.

'This is impossible I'm going back to the office. Let them do the legwork. Why doesn't their map show which ones are free?'

Isabella set off on her scooter and came back to take us to a suitable pitch. It was on row F and too small. We both had a headache.

I drove Katie and Boris round to row F and we started unhitching Boris.

'What is that you have?' a Belgian camper enquired of me.

'We call it an A-Frame; it's for towing the car.'

'It's illegal.'

'For Belgians maybe, but not for us.'

'It's not safe.'

'Thousands of Americans use them. I am unaware of any causing an accident and ours has been fine.'

'Your insurance will be invalid.'

'Bugger off!' was on the tip of my tongue and the A-frame about to be dropped over his head when Gail stepped in and diplomatically explained we were tired and would like to be left alone.

We parked diagonally across the pitch but still into the road; a watching Dutch camper said we could put Boris at the end of the road because no one ever parked there and it flooded. We walked along the beach at the end of the site, had dinner and collapsed into bed.

The next day poor Gail still had a hummer of a migraine. The pitch we were on had to be vacated in a few days because the whole of road E and F was reserved for 40 Dutch caravanners and motorhomers coming for a rally. I decided to look for another campsite while Gail recovered.

I drove Boris onto the motorway heading west, mountains to the right, plastic to the left. The motorway stopped at Adra and the road started to wind round the cliff tops. By Calahonda it had flattened out and I made my way down to the beach and Camping Don Cactus via a mile-long road flanked on both sides by plastic greenhouses. What a tight-knit community. Each caravanner could touch their neighbours. The campsite roads were narrow and the wind was whipping across the site. I found nothing at Almunecar or La Herradura; a Brit I spoke to had to take 3 pitches to cope with his caravan. At Torre del Mar I found my way down to the sea front and 'Camping Torre del Mar'. It was cramped and there seemed to be a distribution of nationalities by row, Germans occupied all the largest pitches in one row, saving them for their buddies when they came available. I despaired of finding a pitch big enough and nearly decided to miss the adjacent site 'Camping Laguna Playa' but having driven over 100 miles reasoned that another 50 yards wasn't going to be too difficult. I found a super 39 ft x 25 ft pitch vacated that morning. I asked some Brits about the campsite – they liked it much better than the one next door. I decided to reserve the pitch for two days and offered to pay immediately – they wouldn't take the money but would reserve the pitch.

By the time I returned to Almerimar, I had spent 8 hours and driven more than 210 miles looking for a site. Gail was still recovering but whilst I was away had chatted to Roy who owned a Holiday Rambler Endeavour. He thought a large serviced pitch was coming available near him. Later I asked in Reception how to get a serviced pitch and was told 'wait until February' – so hope was dashed. We shouldn't have left Villajoyosa.

It was our wedding anniversary but more urgent matters diverted us – what were we going to do? I spotted Roy walking his two Alsatian dogs – he was still convinced that people were vacating the pitch in front of him. The Brits liked to keep together as it were and he was sure that as the large serviced pitches became vacant others had occupied them. I went to Reception to chat to the girl – she said it wasn't possible.

Later I saw her walking the pitches with a checklist and showed her how Katie was too big for the existing pitch and we had been told others just moved on to vacant serviced pitches. She referred us to Isabella in the office who seemed to hold the key to the situation and said we could. We reserved pitch 5003. I came back and gave the thumbs up to Roy and Cindy as they went out on their Harley Davidson, then chatted to the German couple on the pitch who were helpful – the wife had to go back home for medical reasons.

With the sun shining and the future looking brighter, we walked along the beach into the modern port of Almerimar, selected a restaurant on the basis of being packed with local workers and they would know best, ordered *'Menu del Dia'* and must have been served the remains of a sheep that, judging by the amount of fat, had never walked anywhere in its long life but it may have had more to do with me complaining that the soup was cold.

The next day I was up early thinking the Germans in their disciplined and organised way might do the same. They had; the pitch was empty. I raced back; drove Boris onto the large pitch to secure it then fetched Katie. We were on; the sun shone, perhaps we would enjoy it after all but did we really want to be on a campsite with 999 other caravanners and motorhomers, wasn't this Butlins on wheels?

Being on a serviced pitch meant electricity, water and waste were available on an individual basis. Electricity was no problem. The satellite received all the BBC and ITV channels along with Sky news and Price Drop TV. The waste needed a connection from Katie's outlet to the drainpipe in the corner of the pitch. No matter how I positioned Katie, nose in, rear in, slanting left to right, right to left, wriggle her all about, the waste pipe we had purchased in France was too short to reach the drain. A big attraction for men is another man crawling on his back under a vehicle and soon I had copious advice from Roy, Michael, Roger, Jim and Jeff and other unknown passers-by but the laws of physics weren't to be breached. Two of them fetched their spare hoses for me. Using other peoples shit pipes is unsettling. The water tap required another unusual connector so I went to the camp office to enquire if they could supply one. Isabella squawked something into her Motorola and asked me to be back on my pitch in five minutes when a little white van of Juans and Josés turned up, examined what was needed, and set about connecting the water then the waste. What service, Katie wouldn't have to be moved to take on water or empty the waste. I carried the loaned shit pipes back to the campers' vans where, wearing a fetching pair of pink marigolds, I was introduced to the wives, children, grandmothers and anyone else in the neighbourhood.

After a few days of settlement and visits to Roquetas de Mar via a sub-colony of Moroccan agricultural workers and Almeria's seafront market for fake watches and sunglasses, we decided that some fresh mountain air was called for. We set off west in Boris along the motorway shrouded in a heavy damp mist to La Rabita where we turned inland through Albunol, climbed above the cloud to get ever-improving mountain views before reaching Albondon. As we looked back from the *'Mirador'* we could see the cloud formation blanketing the coast and merging with the sea but behind us was the magnificent snow-capped Sierra Nevada and around us hills covered in pink-blossomed almond trees. We felt cleansed.

As we drove over the top the views of the Sierra Nevada were breathtaking. We took the A348 into the eastern Alpujarra mountain range with wild scrubland country to the right and a few more blossom trees on the hills to the left. As we progressed, an almost dried-up river valley stretched into the distance below us with a few isolated stone dwellings down the bottom or clinging precipitously to the sides then a small white village of clustered houses built on top of each other would appear. We passed old ladies walking along the remote road doubled-up over their walking sticks apparently having come from nowhere at all. The road eventually wound into the small town of Torviscon, old stone houses nestling into the ravine carved by a waterfalling river. Old men sat out on the bridge watching a pig trot up and down the main street but nobody seemed to care (*later we learnt this was the town pig fed by everybody and eaten by everybody when the time came*). From Torviscon the road snaked down to a bridge over the Rio Guadalfeo after which we found Camping Orgiva. This small site lying below the steep mountains of the Contraviesa had wonderful views of the snow-capped Sierras. Only a few campers had pitched to enjoy the views of the mountains and the spotless washrooms and there was no sign of Peter and Jenny who had emailed to say they might stay there. This was a Spain we hadn't seen previously and we loved it but could we get the Katie and Boris combination along those vertiginous mountain roads? Anyway having secured our serviced pitch we had committed to a two-week stay.

Without much luck in the dining stakes we gave the camp restaurant a go. The bar was packed with locals but we were asked to sit in the formal restaurant where in splendid isolation we enjoyed a delightful meal with a sun-filled mountain outlook. Replete, we took the winding road through the tunnel along the mountainside to join the main Motril-Granada route at the Rules dam construction site then down to the coast, a quick look around the fortress town of Salobrena and back into the mist.

The Almerimar campsite provided many community activities. I went to a 1-euro Spanish lesson for three hundred people of all nationalities given by the owner's wife Sylvie. I learnt the difference between '*La Policia*' (a light going over) and '*La Guardi*a' (thorough beating) and the useful '*Gorlas en la niebla*' (Gorillas in the mist) although I couldn't remember which branch of the police that referred to. Like-minded nationals used the meeting rooms for social get-togethers for clog making, leather short-trouser repairing, beer drinking and other essential skills. The laundry ran on a 1000 knickers-a-day full industrial scale and the medical centre provided a resident doctor. Gail went to the gym and thought she was in a meditation class but found it was really body-pump for senior campers. Workmen kept the site neat and we regularly returned to find our hedges clipped and the electric cut off.

We went to El Ejido, a modern white-painted concrete town up and over the cliff from us and founded on tomato revenue. We took Boris to visit the local Mercedes hospital for a check-up and service. It looked like every other Mercedes garage the world over. The receptionist spoke little English but came out to say hello to Boris and seeing his parlous state booked him in for the next day although he had already told us he had no vacancies for a week. I cleaned Boris ready for his examination. The next day the receptionist spoke less English than the first one. I took him out to introduce him to Boris and told him they weren't to play with his appendage (brake cable) at the front. He asked for my phone number and I realised I couldn't count beyond five in Spanish so he took my phone and phoned himself. The number came up on his screen and we had donated another £1 to Vodaphone. They called a taxi for me. The Spanish equivalent of Mr Grumpy turned up with no meter and drove me back for a rip-off 12 euros. I had hardly had my first sip of coffee when the phone rang to say Boris was ready so I asked the campsite office to call me a taxi. Mr Grumpy turned up again. Was he the only taxi driver available that day? When we arrived at the garage, I gave him 11euros to test his conscience but he insisted on the 12 euros.

Boris was in fine health but the fridge in Katie wasn't. Ever since we moved it wouldn't work on mains electric so we had to switch to the gas.

Two heads are better than one, or, in my case, any other one will do, so I welcomed Roy's intervention. With his 110-volt heater we found no power to the fridge socket but it went like hell when plugged into one that looked like 110-volt but was 230-volt. Gail reported one trip switch 'was in a different position' and 'feels funny.' With all this expertise it hardly seemed necessary for Michael to come over to show off with his multi-meter and *'Be an Electrician'* kit. He confirmed the socket was dead, forced the trip switch back to its original position and blew the campground electrics.

The next day, and under Nigel's international telephone guidance, we established the fridge was OK but the trip probably wasn't so we should pinch a red wire from there and connect it to the microwave trip. Because of the high correlation between screwdrivers, electrics, spark generation and me as well as wanting to keep a functioning microwave, we decided to keep things as they were. That evening I studied the fridge technical handbook *'Do not use butane'*.

I went to a number of stores but couldn't get a regulator for a propane bottle – we had never seen one.

'I saw one in the camp reception this morning', said our neighbour Jeff leaning over the neatly-clipped hedge.

I raced over in Boris to see it still in the glass sales case.

'It's mine. I'll take it,' I yelled out.

Quietly queuing campers walked over to see what they had missed.

I still needed various screws and connector pipes. Jeff had acquired his from a shop in El Ejido. I parked where Jeff had said but found only a pet shop. I walked past three times disbelieving they would sell propane bits when the shop owner came out, saw the regulator in my hand and pointed across the other side of the road to a *'Ferreteria'*. Bloody obvious, it wouldn't be in the pet shop but the ferret shop.

It was an ironmonger, dark, with a dozen people waiting all squashed up between the door and the wooden counter so that when the doorbell rang and someone else came in they squashed everybody even tighter and those at the front nearly kissed the old lady behind the counter. I showed her the regulator and did lots of miming to indicate pipes and clamps and screw things. She consulted everyone else in the shop and someone else in the blackness beyond the counter and after twenty minutes returned with a collection of bits.

I went to the CEPSA petrol station and tried to persuade the attendant that he was onto a good thing if he swapped my new shiny aluminium Butane for a dirty grey Propane cylinder – he wouldn't because a new legal agreement was required (*they are supposed to inspect your installation*). The owner came and said a payment was necessary or that was my interpretation but did he mean a charge or a bribe? I looked as blank and stupid as possible. He then said something like 'f*** it' but in Spanish, of course, and I walked away with a prized propane cylinder.

I fitted it all together and turned the gas on – the pipe swelled up like a balloon and the connection leaked, otherwise it was perfect.

In Roquetas de Mar I played the high-speed game of ignore the directional arrows to park at the Gran Plaza and *'Bricolage'* to buy a properly crimped pipe, made in France. Unfortunately the French had printed the 'LPG' backwards. I returned to find Gail panicking because I had turned off the propane to the fridge before I left. The next day, with the help of our neighbour Jeff, everything was fitted. We wiped his blood off the piping and thought ourselves lucky he only had a one-inch cut from the Stanley knife because when he fitted his own he had gouged his wrist and halfway up his arm, requiring ambulance attendance and hospital stitching. The gas worked perfectly with no pipe ballooning or leaks. We no longer need to worry about gas in Spain provided we could get bottles of '*propano*'.

After the few days of sea mist, the weather was brilliant with clear blue skies and soaring temperatures. To our rear a sun-bronzed, shaven-headed Roy sat out in the sun with a bottle of San Miguel smoking a fat cigar, his two Alsatian dogs Susie and her brother Caesar at his feet. You wouldn't want to tangle with the dogs or Roy either. He, and wife Cindy, had sold up everything and their 38ft blue Holiday Rambler Endeavour was now their home. It had a huge tent-like awning on the side to give them an outside room. Roy told me that they were keen motorcyclists so they also had a Harley Davidson motorcycle and a bright yellow Fiat car, both of which were towed on a trailer that he had at the rear of the pitch. After he had told me the story of buying their RV (more dealer grief) he proudly showed me around. In the lounge he had two white leather three-seater sofas, furniture in light maple and a large kitchen with a tiled floor. The bedroom was the same as Katie's but the bathroom between was open plan, a concept I had always found difficult because you walked past the open bog on the way to the bedroom. We invited Cindy and Roy over to Katie and perhaps to compensate for her more modest accommodation I spent a great deal of time showing them how to open and close the toilet door.

January 16 was Gail's birthday and the sun shone nicely for her. She called her parents as a pre-emptive strike against their call to her. We went to the port in Almerimar for lunch and sat peacefully outside in the sun alongside the moored yachts to finish our coffee.

'Let me down, let me down!' a tenor voice called out.

High up on one of the yachts dangled a middle-aged man in khaki shorts, like a baby in a baby walker, his weight supported solely by a cloth strap under his crotch that had pulled his shorts up and threatened to expose his manhood. The bosun's chair he had been hoisted up in had slid round and was up his back. He vainly grappled to pull himself up on the suspending rope to ease the excruciating pressure on his vital anatomy, his face reddening with every moment.

'Lower me down!' he pleaded to his wife who stood paralysed on the boat deck below. She leant down to grab something and he plummeted towards the deck to join her.

'Not so fast!'

Suddenly he came to a jarring, testicle-wrenching stop, dancing like a puppet on a string, tears visible on his face. I unconsciously shifted in my seating position.

'For Christ's sake, woman!'

He was released and once more plunged jerkily to the deck under a full gravitational force crashing in a heap. He cautiously stood up and saw us watching.

'I'm going to be impudent,' he gasped.

A Freudian slip but we had the picture.

Without speaking to his wife, he stormed off the boat to the table next to us and ordered a beer, sat down and with a sharp intake of breath immediately stood up again. What can you say to a man in that position? 'My wife Gail is good at first aid,' so 'would you like her to look at it for you?' didn't seem right. Eventually his wife came over but the air was icy. We had a chat. He never spoke to her and she contradicted everything he said. He was British, his wife Norwegian, they lived in Norway and spent six months cruising Mediterranean ports.

We headed back in the sunshine along a near complete marble, palm-planted, ornately-lamped promenade; behind which stood new blocks of flats and in front a deserted beach with smart shower emplacements at regular intervals. Earthmovers raced each other inside the red and white tape markers to fill in the few remaining gaps. The promenade stopped where the campsite jutted into the beach then continued the other side way into the distance waiting for its hotels and blocks of flats to be built. At the far end were motorhomers' free camping including some Germans we spoke to in a magnificent Monaco motorhome that had probably cost them £250,000 and left them with no money to pay the £10 a night for secure camping.

Gail's sister and then her brother called on the mobile, further swelling the profits of Vodaphone's shareholders. The campsite Reception had received a parcel containing six kilograms of mail forwarded by the management company at our apartment. It was like Christmas, except we had paid £70 for the delivery. We sat out in shorts and T-shirts opening bill after bill including the monthly one from Vodaphone for £256 but we still had no satnav discs from Nigel.

On January 20 we drove east along the coast through Aguadulce and Almeria to the mountain village of Nijar, the streets lined with people, pots, baskets and straw donkeys and found we had inadvertently become part of procession that eventually finished at the cemetery. After we had paid our respects, we extricated ourselves and drove through the brown plastic-free hills of the National Park to the Puerto at San José where we parked opposite a row of sun-drenched restaurants. Hearing British accents at one end, I asked whether that was the one to eat at and the owner replied that it was. I noticed he had a fulsome sandwich so ordered the same for both of us. As we ate our sandwiches the owners of the restaurant told us things were going well for them particularly now they had a Greek chef and they loved the place. The Spanish were friendly and his wife told us how her daughter had fallen off her bike and a complete stranger had brought her home safely. I reflected how unlikely this would be in the UK, the stranger almost certainly being arrested for kidnap and paedophilia. As we walked back to Boris, Gail told me there were no sandwiches on the menu, only main meals, the owner was simply having his own lunch. We discovered it was impossible to drive over the cliff to Cabo de Gata from San José although the map showed you could. Impressed by postcard pictures of craggy rocks and waves crashing over an isolated lighthouse, we took a long diversion, drove across the salt flats, up the narrow rocky peninsula road and down to the lighthouse and wished we'd never bothered.

That evening a party was held in one of the campsite clubrooms for Peter and Michael who would be leaving in their American RVs on Saturday January 22 to go to Marbella. We were entertained with clarinet from Roger and saxophone from Jim, but it was the howling accompaniment of Whisky a brown and white terrier dog, that stole the show.

The next morning the camp electrics were out – had they been cutting the hedges in the night? Thank goodness the propane was hooked up so we had the fridge running, super heat, a kettle boiling for breakfast and hot water for showers.

An elderly German couple moved to the empty pitch next to us and to pre-empt any John Cleese thoughts, I helped with manoeuvring their large caravan. It took them another week to erect their awning, kitchen tent, cooking equipment, outside fridge, windbreaks, dining table, barbecue, chairs, loungers, parasol and dig their defensive ditches and string the barbed wire.

'Will you be staying long?' I tentatively enquired but the attempt at humour was lost in translation. I went down to the camp office and reserved our pitch to the end of the month in case they had ideas on further expansion.

Saturday arrived and Michael's diesel Monaco was purring nicely ready for the off. Peter's RV was stranded like a beached whale. It had been driven forward to make a turn across the camp road but became lodged against the hedge opposite and then refused to go either forwards or backwards. The engine was willing and Peter put on a fine display to show it would rev all the way up to the 5,500 red mark without exploding but the wheels wouldn't turn. Twelve men, all hoping to appear super intelligent by pointing out he had done something stupid like not having it in reverse gear, peered in the cab, then under the bonnet, at the wheels, under the wheels and into the sky but no salvation was forthcoming. The campsite now concerned that one of its roads was blocked arrived with its rubbish dumper and tried to tow the RV backwards onto the pitch. It was immoveable. Wheels were jacked up and still wouldn't turn. After asking whether he had left the brake on, my mechanical knowledge was exhausted so we continued on our planned drive but I did think about what might happen to us in the same circumstances.

We drove through Almeria then north to Tabernas and through the desert area known as Mini-Hollywood, where I envisaged Clint Eastwood and *Laurence of Arabia* scenarios, to Sorbas. A dramatic scene of a white-painted village perched on a cliff lay before us and we were pleasantly surprised to find an open information centre within. Town map in hand we did the suggested walk, but were unable to find the Castillo, and the mighty Rio de Aguas had dried up so felt cheated and retired for *tapas* to a typical Spanish cobbled town square and a building called '*Pub*'. After lunch we took a small road to Los Molinos Del Rio Aguas with wonderful views back towards Sorbas then onto the motorway to Garrucha and Mojacar where we hunted and found Camping Cueva Negra. Here we surprised our friends Peter, Jenny and Ben the dog and spent a hot sunny afternoon socialising. Their campsite was new, small, nestled in the hills, beautifully planted and maintained with all pitches fully serviced. I invited them to Almerimar 'you really must see it.' The coastal views on the return journey through Carboneras were spectacular. Pulling up alongside Katie we saw that Peter's RV was back on the pitch.

Early Sunday, Michael was warming his engine again and Peter pulled out to attach his trailer.

'What was the trouble?' I asked

'The parking brake was locked on. Michael had to dismantle the cowling where it works on the propeller shaft and release the shoes. Just have to do without it.'

'Thought it must have been something to do with the brake,' I said, resisting the urge to say I told you so. The usual suspects gathered to watch events and as they all waved their final goodbyes, I noticed Peter's bonnet wasn't closed. Should I tell him when they had been dismissive of my earlier technical advice? I did, he was grateful and I hoped nothing more went wrong for them. Two hours later after I had been to Almerimar port to get the papers and done some shopping, I saw them at the filling station in El Ejido – would they ever get going? They had become more Spanish than the Spanish.

On Monday January 27 we woke from a lot of wind in the night, unrelated to any gastric problems. It continued as Peter, Jenny and Ben arrived for their day with us. The normally calm sea was pitching Atlantic-type rollers so we abandoned a beach walk and drove to the port for lunch. On our return a queue of fifty Dutch motorhomes waited outside the campsite, so I was glad we had booked in to the end of the month. After our visitors had left I mentioned to Gail I was getting a cold.

'No you're not, think positive; it's all in your mind.'

'It's all in my nose.'

We slept little because of continuing winds and I *had* developed a cold and after breakfast to prove it I crawled back to bed and didn't emerge for four days. On the fifth night I hallucinated about internal body organs that had gone wrong ending up in El Ejido hospital where they strung me up with a harness between my legs undecided whether to castrate me and I couldn't explain because I hadn't moved onto tape 2 of the BBC's Spanish in a week.

Later in the day I emerged and sat out in the healing sun. A newly-arrived German camper opposite unusually had his van radio on very loud. After the tenth continuous playing of '*Itsi, bitsi, teeny, weeny, yellow polka dot bikini*' I was ready to start World War Three and asked him to give it a rest, and he did. The German lady from next door waved and came over to talk which was limiting as she didn't speak any English and I had exhausted my German on Mr *Itsy Bitsy*. She said, '*Zwei*' something or other. You are complete, two again; she had noticed I had been missing for the past few days.

Gail had done everything, including dumping the tanks. Our good neighbour Jeff had fetched a newspaper for me every day. He, wife Milly and mother-in-law Rose seeing me out again, all stopped to chat. They were in a massive American fifth-wheel that had been towed down to the site by Jeff's brother and left permanently. John from across the road asked after me. It was his first trip with his wife and two delightful Westies in their RV and he had found the whole thing a bit daunting. Pedro, a free-spirited and ugly little brown dog visited as he did every day on his way down the forty roads from the stables to the beach and back. He would say '*Buenas Dias*' to all the other dogs, particularly the females all restrained by leads and owners desperately trying to get rid of him but he was an old hand and knew the game. As I sat out and recuperated I was constantly amazed by the number of couples walking the site in matching tracksuit or anorak tops and mentioned this to Gail thinking we should be more fashionable ourselves but received a dismissive response.

The next day, for the first time since we had arrived, a light dusting of rain fell on the window but then it had gone. I told the campsite we would stay beyond the thirty days and they demanded 200 euros immediately for the electric that would now be charged for rather than being inclusive but at a reduced rate, I was confused.

Sunday February 8ᵗ and fully recovered from my cold, we decided on a trip to blow away the cobwebs. We set off for the mountains via Berja, to Laujar de Andrax and a narrow secondary road where we took in impressive views of the snow-capped Sierra Nevada. Near Bayarcal Spanish families took full advantage of a delightful public picnic spot's barbecue grills the amount of smoke suggesting they were cooking enough food for the whole of Andalucia. We climbed to the summit at Puerto de la Ragua. We had come from a scorching beachside to cold deep snow in no time at all. Over the top of the pass even more spectacular views met us. As we descended farther into the foothills we saw, looming over the village of La Calahorra and the plateau of the Marquesado, the stunning redstone Castillo de La Calahorra. Apparently it was one of the first Italian Renaissance castles outside of Italy and constructed between 1509 and 1512 on the site of a former Moorish fortification. It stood on a small hill, its towers resembled four huge pepper pots. We took the motorway to Guadix to say hello to the Troglodyte cave dwellers; they were quite normal and didn't have stoopy backs from all the bending over and within the town we discovered an imposing castle and cathedral.

We had fallen in love with the Alpujarra mountains and the following Sunday set off on an extensive tour along the north western rim from Alcolea, through Laroles, Valor, Yegen (of Gerald Brenan fame) Mecina Bomberon, Berchules and eventually Trevelez where the Queen buys her ham and the Spanish had invaded for Sunday lunch. The vistas were stunning and became even better as we continued up to the tiny white villages of Bubion and Capileira stacked against the lower slopes of the Sierra Nevada trying not to spill over the cliff faces. Again both villages were packed with visitors. We drove out of Capileira up a twisty unsurfaced mountain road to 7,000 feet to enjoy spectacular views of hazy-blue mountains that went almost to the shores of North Africa then descended west through Orgiva before returning to the coast.

Michael, Peter and half-a-dozen other Brits had already left, our neighbours Jeff, Milly and Rose had flown back to the UK, two whole streets of Dutch campers had pulled out en-masse, the sunshine was continuous, the temperature constant at 24 deg C and sunbathing had become passé. I had even completed tape 2 of the language course. Then on February 17 after months of waiting, our satnav discs arrived along with two new waste pipes to replace our French 'shorty'. Peter and Jenny e-mailed, they had moved to Ronda. It was time to go. We went to pay the bill and spent the next hour arguing over it.

Granada and Fuente de Piedra

'What's fully charged and makes your blood boil' I shouted from within the toilet.

'I've no idea – will you hurry up, we're supposed to be leaving for Granada this morning,' pleaded Gail from the bathroom.

'It's an electricity bill from this campsite. If you stay a month they include the electric in the daily charge. If you stay for a month and a day they charge you for the electric *from the day you arrived*. Bloody mad!'

I opened the toilet door just as Gail stood on one leg to put her jeans on, and sent her sprawling on the bathroom floor.

'Keith, will you go outside and get Katie ready.'

It was a little colder. We soon had Katie readied and Boris hooked on.

234

I went to the campsite supermarket to buy some treats but as I wandered back something was nagging at me to re-check Boris. I opened his door. The keys were in and turned to the right position, the radio off, heater fan off, the brake cable attached and a bungee stretched from the pedal to the seat support to make sure the pedal returned to the off position. I checked the handbrake.

'We left Boris's handbrake on.'

'Keith, that has always been your job.'

A cold sweat passed over me as I pictured Katie pulling Boris with his brakes hard on. I had read of American RV owners dragging their cars for miles with the brakes left on and being waved down by passing motorists unable to see through the smoke clouds from burning brakes and shredded tyres.

We drove to the reception where I presented gooey chocolate cake and biscuits to the girls. Perhaps 15 euros a night wasn't so bad really and they weren't responsible for the charges.

We drove up the cliff and onto the plateau where Katie's satnav proudly announced itself back in action. We dropped onto the motorway and travelled east towards Almeria then passed the N340 to join a new motorway at junction 453.

'Please make a U-turn.'

'What? We've waited five months for the discs, paid a hundred and twenty pounds and another fifty pounds for delivery, driven forty miles and it's lost,' I moaned.

Katie's new Spanish satnav disc denied all knowledge of the motorway and as we turned north the monitor showed us off-road, tracking across a desert. We had driven the road before on the way to Tabernas and remained fascinated by the drama of that surreal lunar landscape. Dry riverbeds, eroded gulches and deep ravines interspersed the barren desert slopes, yet we were on a super modern highway a few miles from the metropolis of Almeria and the Mediterranean. It was an inland sea of sand without vegetation, habitation or thankfully plastic.

We joined the A92 and the satnav recalculated the route.

'Isn't it brilliant how it does that?'

'You've hated it for the last fifteen minutes.'

The road continued to be as much a desert as the landscape. Katie worked hard to climb the hills and after two hours of driving and an unchanging vista we stopped on one of the few service stations for a quick bite to eat in Katie. Ten minutes later the *Guardia Civil* arrived to circle us in their green and white car like vultures eyeing up a stricken prey. We sat paralysed and they eventually drove away. Relieved, we set off only to be joined by two crisply uniformed, Ray-Ban-shaded *La Policia* on motorcycles. It was ten minutes before they turned off to Purullena. I had concentrated so much on our followers that I nearly missed the contrasts in the scenery. Off to the left the northern peaks of the magnificent snow-capped Sierra Nevada had appeared whilst down at road level we traversed a desert landscape of dusty ochre soil and hills punctured by the rabbit warren of caves of the Guadix and Purullena Troglodytes. The road climbed then eventually fell steeply towards Granada. We circled to the west then south, followed signs to *'Sierra Nevada'*, and turned off to *'La Zubia.'*

The campsite had a lovely entrance behind a pull-in road and a restaurant on the right. Inside it was a delight, like a walled garden with a tree-lined rectangle of concrete roads and a central avenue bisecting the site. It had twenty attractive little log cabins on the right, and a facilities building at the far end with a swimming pool above and forty-six pitches. What a complete contrast to Almerimar.

Once inside, Katie looked enormous and we realised she was too big for the assigned pitch and the one in the corner against the back wall was the only one that would take her. We settled in, the TV worked, tea was brewed; and the sun was shining in.

The next day I found the washrooms, not that we needed them, were spotless and heated as were the laundry rooms. The trees were luxuriant; it had a shop, phone, and the restaurant looked tempting. For 0.80 euros you could take the bus from the road outside the 2 miles into Granada.

For those expecting a detailed description of a tour round the Alhambra disappointment reigns because we had been before so gave it a miss. Instead we spent the first day re-familiarising ourselves with the centre and avoiding the rain; the first we had experienced for months. On day two we went in by bus and as we descended at the bus terminal two guys accosted us asking for money. I engaged them in a philosophical discussion about how they were better dressed than we were with smart new gold Nike trainers and tracksuits, so maybe they should think about giving us money, when Gail linked arms and hurried me away.

We walked up to Puerta Real (a square), Plaza Nueva and Plaza Sta. Ana to a narrow cobbled road running beside the River Darro with the fairy-tale red towers of the Alhambra Palace dominating the cypress-wooded slopes above, and spent thirty minutes trying to get a winning photo like the millions before us. We went over rivers, up little cobbled streets, up and down hills, along white-washed terraces, on ancient city walls, peered into cavernous Flamenco bars being swilled out from the night before, realised that miradors provided better photo opportunities, and squeezed through a cobbled square busy with markets stalls. We had coffee in a bar nestled in the city wall and overflowing with people, imagined gypsies living in the cave houses, saw cubes of Moorish white-painted houses, passed ornate churches and monasteries and then were rained on and buffeted by the wind. Three hours later we emerged into a narrow alley and a Moorish teahouse for another coffee and cake with our legs resting on the velvet stools having done the atmospheric labyrinthine Albaicin and Sacromonte districts.

Buoyed up by a meal in the student quarter, we visited the cathedral where photos were allowed and then Capilla Real the last resting place of Isabel I and Ferdinand V, known as the Catholic Kings, where they were not.

A rainbow of colours was on show as we woke early on Sunday. Crowds of Spanish weekenders wearing bright ski suits were standing around enjoying a last coffee before they set off for the slopes. Mothers tried to persuade children to get dressed or use the bathroom one last time whilst everybody questioned everyone else about their plans, their equipment, their new car, their grandmother, until eventually some force drew them into their mini-buses or cars but only after someone jumped out again to check something for the last time. As it was a bright start to the day we decided to follow on after breakfast.

The A395 was close to the campsite, busy, and took us steeply up the winding road into the Sierra Nevada and it wasn't long before the snow-capped jagged peaks came into view and then snow was all around. Since we have never been skiers but are always anxious to explore as far as we can, we kept driving until the road ran out way above the hotels and ski lodges to a parking area cleared of snow adjacent to a series of wooden huts selling all manner of snow equipment and souvenirs. Brightly clothed children were hiring sleds or snowboards to career down the slopes to crash joyfully into groups of chatting parents or innocent passers-by. We climbed to a viewpoint on the edge of the mountain and marvelled that all this was only a two-hour drive from the Mediterranean. We quickly became exceptionally cold and opted for coffees standing at a counter in one of the little wooden cabins. That coffee was so warming I could trace its path down my throat and body but then realised it was rapidly heading for the bladder.

'Where's the toilet?' I wondered aloud. None was obvious.

'Perhaps it's round the back,' said Gail. But it wasn't.

'There has to be one somewhere for all these people.'

'Perhaps they can't have them because the pipes will freeze up,' suggested Gail.

We looked everywhere, the warming sensation now becoming quite pressing.

'Dónde están los servicios?' I asked a stallholder but he shrugged his shoulders then laughed.

'Do skiers do it in their ski-pants like surfers in their wetsuits?' Gail suggested again.

'Frankly I don't care – we're going.'

I slalomed Boris down the mountain like he was on skis and skidded into the El Dornajo visitor centre halfway down and raced for relief.

'What great views of the Genil valley and Güéjar Sierra from here, I'm so glad we stopped,' I joked afterwards.

We drove farther down and across the neck of the lake (Embalse de Canales) and up the other side where we had a picnic lunch at the viewpoint, Güéjar Sierra above us Granada below. It was a wonderfully sunny day but that evening it rained again, the weather had changed since we left Almerimar.

Two British guys were on the pitch next to us in a VW camper. The next morning I tried to make friends with their dog, convinced it was about to pee on my expensive stainless steel wheels but it snarled at me.

'Where are you guys headed?' I enquired.

'Portugal, we should be there by three this afternoon.'

'I keep hearing about motorhomers getting broken into. I guess you guys don't get any problems having the dog?'

'Oh he's as soft as they come. No, we carry CS gas. We buy it in France but you should be able to get it in a hardware store in Spain.'

We walked to La Zubia.

'Buenos Dias. Tiene CS gas anti robo?' I enquired confidently of the proprietor of a dingy *ferreteria* after squeezing past other customers to scour his many shelves of everything you thought never really existed.

My Spanish must have impressed them and it was a moment to appreciate – silence whilst amongst the Spanish. Heads turned in our direction and the proprietor seemed to struggle for the right words but only uttered the one.

'Granada.'

I thanked him and as we left turned back to wave at the faces pressed against the window. We went to Granada but didn't see any. We would simply have to take our chances.

The next day we drove Katie and Boris out of the delightful Reina Isobel campsite to find the *Guardia Civil* in two cars waiting outside. They looked at us in only the way the police do as you drive past but is was probably more to do with never having seen anything like Katie and Boris. We drove through Granada, enjoyed more views of the Sierra Nevada then headed west to turn northwest after Antequerra and an uneventful 80 miles to Fuente de Piedra.

After traversing the village, we spotted the campsite by its many colourful national flags sodden by the heavy rain and a huge '*Camping*' sign over the top of the gate accompanied by a height restriction sign for 3.5 metres on the gate pillar. As Katie is 3.65 metres, it was a matter of some concern.

'It looks higher to me but we'll just have to go for it we've no other choice,' I said to Gail.

Gail disembarked and watched me in. Aerials rattled but we were otherwise all right – they must have rounded the figure down or, it was the only sign they could get.

Camping La Laguna was small and we had to take two of their sideways sloping gravel pitches. With the rain teeming down, none of the other four British campers was going to step outside to socialise. Despite being small, the campsite had bungalows, a restaurant, swimming pool and Jacuzzi.

The lake that made Fuente de Piedra famous was just visible through the rain streaming down Katie's window. At 1.5 miles wide and 4 miles long, this shallow salt-water lake became the favourite stopping-off point for a reported 16,000 pairs of flamingos that wanted to breed inland and were tired of the crowded flamingo waters in the Camargue and as a result became one of the most densely populated bird habitats in the world. We drove Boris the 9 miles around the misty lake, commenting on some smart houses amongst the olive groves and were fortunate to see flamingos wading in the water and scouring the bottom for food but it wasn't the pink heaven we had hoped for. The observatory was closed and would open the next day. We could just make out the island where the flamingos breed in spring. Whilst sitting there they apparently worry that the water might dry out so fast that predators can get across from the mainland. I also read that in Roman times they were hunted for their tongues, a delicacy apparently. The village itself, like all places after 2 p.m., was closed, but we admired the tiled pavements lined with fruit-laden orange trees and their trunks painted white. By now the rain had resolutely set in so we returned to Katie, a damp evening and no TV.

Arcos de la Frontera

It had rained in the night. So in preparation for moving I had to remove the water that collected on the slide-out awning. I did this by climbing a ladder leant against the slide and scooping the water with a jug into a bucket then, depending on the proximity of people or motorhomes, chucking it or descending to pour it away somewhere. I had the bright idea of using the long rod that we used to reach the loop on the main awning. By pulling down slightly on one end of the slide awning the water might run off of its own accord. The lightest touch was enough. The water ran off the awning down the rod, down my arm inside my anorak and shirt. It was cold and surprisingly wet. As the sun came out I began to steam.

We paid 15.75 euros for our stay. I persuaded the pleasant young girl in Reception to phone our next campsite at Arcos de la Frontera to check everything would be all

243

right. We were going to Arcos because I couldn't resist the evocative name, imagining an old wild frontier town; it also claimed to be one of Andalusia's most dramatic white villages perched on a cliff above a river.

We ignored the unmarked road to the right of the campground and went back through Fuente de Piedra to reach the Autovia (A92) then onto the A382 (Jerez road), a single carriageway across the plains between the mountain ranges. The clouds darkened and it started raining heavily with thick spray from other vehicles, so driving required all my concentration. I was thankful to be sitting up high out of the worst of it but amazed at the risks people took to get past us in their cars through our own impenetrable cloud of spray. Then a shaft of light, like a theatre spotlight, pierced the blackened sky onto the hills away to the right highlighting the white village of Olvera and its church. Was it pre-ordained from on high to make sure we didn't miss the beauty of that sparkling village, like a snowcap on the mountain? We were momentarily transfixed.

The worsening weather and increasingly twisty hilly terrain made driving more difficult. When we emerged from a tunnel, just before Algondales, misshapen vehicles and debris were strewn across the road with police and ambulance men peering into them eerily bathed in the blue and amber of their flashing lights. We passed Villamartin and relieved stopped for a fuel top-up and comfort break; an 85-mile journey seemed to be taking forever. Soon after, at the end of a long and straight road far in the distance, we could see Arcos de la Frontera high on a hill.

As we approached the white town we found our left turn at kilometre 29 and the A372 signed 'El Bosque' and 'Camping'. We descended a hill, crossed a dam, took a left turn to 'El Santiscal' and seeing the flags on the right, pulled in through the white concrete gateway and parked alongside the site office and restaurant. The rain continued to bucket down drumming on Katie's roof. Should we even bother to get out, we wondered?

Gail donned her anorak and hood and looked cute; I put on my 1-euro Benidorm-bought poncho and resembled a giant yellow condom. Gail ran to the welcoming office where she was invited to choose any pitch other than those occupied by the other three campers. The site was flat and open like a large grassed garden but many pitches had standing water and on others, when I pressed my foot down, water came over the top of my trainers. What were Katie's 10-tonnes going to do? I was convinced she would sink. In addition almost all pitches had Moncofa-style awning rails that were too low for us to enter. In desperation I considered staying on the tarmacked parking area. The young guy and girl in the office wanted to help and said we could park on the formal lawns near Reception or, next to an American RV, which had been abandoned by its owner. We had already thought how clever of the owners of the RV to select the only viable pitch. We parked in front of the RV next to the flowering but dripping mimosa trees. The electric point was two cable lengths (50 metres) away. We were pleased to settle and the TV worked first time.

Exciting, and at the same time scary thunder and lightning, accompanied by torrential rain, continued throughout the night but by the morning the skies had cleared. After the traumas the day before we were happy to linger over coffees but by lunchtime we were bored and decided to drive in Boris to Cadiz only 45 minutes away along a quiet A393/A390.

The approach bridge was striking and then we saw the deep green of the Atlantic. The long main street through the modern part of Cadiz could have been any European City but as we went through the triumphal eighteenth-century Puerta de Tierra, the gate in the stout defences of the oppressive thick high stone walls, we had a sense of excitement at entering the oldest city in Europe. Emerging you could go left or right of the narrow fortified peninsula that stretches out tongue-like into the Cadiz Bay and the Atlantic. We chose to go left to the south and west and parked on the street opposite the royal prison (Carcel Real), built in 1792 and described as one of the most important Baroque civil buildings in Andalucia. We hoped some old lag didn't break out and steal Boris for a getaway car. I checked with two passing wardens that we could park on the meter but when the ticket was issued I saw it expired at 10.17 a.m. Confused, I showed it to the warden who explained – *'mañana'* – tomorrow – all for 1.2 euros.

We headed through the narrow streets to the 16 C Plaza San Juan de Dios, the nerve centre of the city where the City Hall (1799) and other fine buildings exist as well as the tourist office that was advertising many fiestas. Equipped with brochures and guides we sat at a table in the large open square and had coffee watching the world passing by. After a visit to the *servicios* I mentioned to Gail the wonderful array of *tapas* in the bar, so with threatening skies we stayed for lunch.

The display of *tapas* was overwhelming and beautifully presented, dishes of oysters, squid, clams, langoustine, shrimp, sea snails, sole, hake, cold hams and meats, stews, beans, artichokes and many we couldn't even guess whether they were animal or vegetable and it took some while to make our selection because every time you chose something another one looked more tasty. The locals overcame this problem by sitting at the bar in close proximity to the food so they could ponder their next choice as they ate. We had ours delivered to our table outside and sat under an umbrella the rain clouds passing overhead.

We enthusiastically started our post-prandial tour by leaving the square for the district of El Populo, the medieval heart of the city and thereupon gave up following the guidebook route. We had more fun wandering the narrow cobbled alleys through the three districts or '*barrios*' to see the New Cathedral, the flower market, the gardens in Plaza le Nuna, three ladies in traditional Flamenco dresses chatting in Caldera de la Borca, and the Atlantic Baluarte de la Candelaria Here, embraced by westerly winds we looked out onto the rolling Atlantic breakers and wondered about the trepidation that Columbus's crew must have felt when they did the same. They expressed their relief on return by planting the 'giant' tree garden. Could they ever have imagined us taking silly tree-embracing photos or the sunbathers on the beach below? From the same gardens we saw Alameda de la Apodaco, with the Iglesia then walked through Parque Genoves, a sanctuary of sculpted trees, to arrive at the south side of the peninsula alongside the ocean. From Campo del Sur we looked towards the cathedral. The winds had cleared the clouds from the sky and the golden cupola glinted in the sun above the colour-washed houses on the streets below. There we finished our circular tour and felt we had done justice to this ancient city (founded by Phoenicians around 1100 BC). For once school history had come to life as it was here in 1587 that Sir Francis Drake '*singed the king of Spain's beard*' with a raid on the harbour. By so doing he delayed the Spanish Armada and on October 21 1805 Admiral Villeneuve sailed out of Cadiz to have his fleet destroyed by a mortally wounded Nelson at The Battle of Trafalgar.

We returned unintentionally via the white hill top town of Medina Sidonia but decided against viewing its Roman sewers and crossed the motorway to be eventually surprised by the spectacle of a glistening sunlit Arcos perched on a sheer cliff edge, its white houses tumbling down the hillside.

The next day the bottled gas ran out.

'I saw a CEPSA garage in Arcos as we passed. I'll pop over and get the bottle changed,' I proposed.

I drove in to see a fine display of shiny butane cylinders but they didn't have propane. I returned to Katie and used the satnav to look for CEPSA garages; almost all were in Jerez. I found one on the way but only butane was available and then the same thing happened in Jerez. I had assumed all CEPSA garages supplied propane like the one in Almerimar. I stopped and bravely asked a man for directions but he must have misunderstood and sent me to the Sandeman and Gonzalez Byass sherry estates 10 miles out in the country. Things were desperate; I had to find a toilet soon. I managed to get back into town by driving across some industrial estates but then found myself driving north up the N1V and knew I would soon join the *'Autovia del Sur'* and be heading for Sevilla. At the entrance to the motorway I saw a CEPSA garage that, thank God, was on my side of the divided highway and great good fortune had a stock of grey propane bottles round the back outside the toilets. The propane was 7.20 euros and the cheapest yet. He seemed mystified by my relieved happiness. I returned triumphantly with my cylinder. Gail wondered where the hell I'd been.

I chatted to a British couple in a VW campervan like the one we used to have in America – the hippies as we call them had been to Barcelona, Cabo de Gata and would probably look for work in exchange for food in local agriculture.

On Saturday February 28 we set off for Seville in Boris. The day was clear, as was the road going north. We were surprised to have to pay tolls again (5 euros). 1½ hours later we had parked underground and were a brisk walk away across the canal from the centre of Hercules's city. Open horse-drawn carriages passed us by as we eagerly walked along Paseo de Las Delicias, through Plaza Puerta de Jerez then down Av. de La Constitucion to The Real Alcazar. The images of Sevilla in the guidebooks are so wonderful that we steeled ourselves for disappointment but we found a vibrant beautiful city that exceeded our expectations. We were among many other tourists but it didn't matter, wherever you looked, the wealth of historical buildings presented a photo opportunity and everybody seemed to be smiling. As we entered the Alcazar and the Patio de Banderas, we turned momentarily to be surprised by a splendid view of the cathedral and its magnificent tower La Giralda. Somewhat overwhelmed by it all, we walked aimlessly down the side of the Alcazar into tiny streets with flower-bedecked balconies, the Plaza de la Alianza and the Plaza de Dona Elvira. We pressed our faces to the iron bars of the gates of many of the bijou houses to see their hidden cool marble patio gardens filled with succulent and delicate green foliage plants. We hoped to find our way into the Alcazar but ended up walking the full perimeter of its huge defensive walls wishing we were on the other side to eventually pass the tobacco factory that was and now is the university and back into Calle San Gregorio after which we found the entrance. With the audio guide we were able to explore this magnificent sprawling royal fortified palace, one of the most important examples of Mudéjar architecture. Started in 913 it had many Christian additions right up to Franco's time. It consists of a collection of palaces linked by a maze of corridors, with vast rooms, patios and halls representing architectural styles from Islamic to Neoclassical. Magnificent tapestries hang on the walls and the formal gardens outside were a delight.

Outside the cathedral, one of the largest in Christendom and La Giralda (the minaret that was copied in Rabat and Marrakech) we were spellbound by their grandeur and exquisiteness alongside the joy and bustle of tourists deciding whether to hire a horse-drawn carriage. At La Cuera, an attractive restaurant in a small square near the Alcazar walls in the Santa Cruz district, we sat outside and ate steaming plates of paella. We walked to Plaza Nueva, but were unimpressed by the Town Hall, then to Capilla de San Jose but couldn't find the Palacio de Lebrija and ended up in Plaza del Salvador that was jammed with crowds of wholesome, well-behaved, young people, drinking, eating *tapas* and talking loudly against a backdrop of the pink Iglesia del Salvador.

We headed for the river via the Bull Ring (La Maestranza del Caballeria) then crossed Puente Isabel II o de Triana, leaving the tourists behind. We turned back along Pureza and then diverted to see the church of Santa Ana. Although the restaurants and bars were full here, the atmosphere had changed with local families out to enjoy their weekend, all part of this wonderful city.

The next morning a strong sun rose over the hedgerow and into the 'dining room' as we breakfasted. After two days of intensive tourist activity, Sunday provided an opportunity to relax so Gail did the washing and ironing and I phoned home. The second of our two telephone cards bought at La Cala (Benidorm) finally ran out. What tremendous value these had been. I hoped the new card would work as well. The washing dried on the line as I sat out reading. Two overnighting motorhomes departed leaving only one Spanish family who appeared around 11 a.m. Later, ageing hippies arrived in a small, lime-green Mercedes van with stickers filling the windows showing places they had visited and a tropical sunset scene with palm trees painted on the side.

Although we had been at the campsite for five days, we had not explored Arcos de la Frontera itself. We approached from the east and by so doing acquired a tremendous view of the limestone cliff face towering over the River Guadalete with the town perched on top its various spires and sandstone castle standing proud over the white houses that cascaded down the sides. We crossed the river bridge and ended up at Puetra Matrera – an Arabic wall and tower and fortunately found a parking place in the small collection of houses outside the gate. Walking up the road (Calle Matrera Arriba) to the Calle La Pena Vieja we arrived at the Mirador de Callejas that afforded exhilarating views over the town, the rolling plains and, looking north, the lake, and was that our campground to the right?

We nosily peered behind massive wooden doors and doorways as we went up the pretty and striking cobbled streets of whitewashed houses to see courtyards and colourful flower-planted patios. We juggled two tourist leaflets, one in Spanish with a street map and a list of twenty-five historical sights and another in English with a historical background and information on churches, civil architecture, convents, mansion houses and Andalusian architecture such as the castle and gates. Unfortunately one did not match the other.

We soon reached what, on one map at least, was the Iglesia de San Augustin (1539), an ancient convent where only the church remained, and startled the lady tending the flowers by one of the altars. She led us to see another altar from which, apparently, the figure of Jesus receives fervent adoration at the head of a dawn procession on Good Friday.

On Calle San Pedro a payment of 1 euro allowed entry through an eighteenth-century Baroque façade into the fourteenth-century Iglesia de San Pedro (Peter) built on the site of an Arab fortress. We had visited many memorable churches and cathedrals but here facing us was an absolutely stunning; sharp-intake of breath; glittering; ornate gold side altar. With spiritual music playing, the scene transfixed us. Only after a few minutes of serenity did we realise it was only a taster for the super-bling of the main altar. We left to see that the lady and gentleman who took our money at the door sat engrossed, reading ancient manuscripts, and hardly noticed our departure.

We had kosher coffee in the House of Jesuits, a small market square where I managed to interpret for some visiting Brits who had been rudely shouting 'beer!' in greatly increasing volume and exasperation at the young waitress who didn't understand. Apart from this open space, historic buildings and palaces lined every inch of the narrow streets.

The convent of the closed order of Mecedarian nuns had been a prison until 1642. We stepped beyond the giant prison doors into a vestibule where in the wall was a small rotating wooden door at head height, two electric bells and a cord pull for a hand-rung one. We rang tentatively and eventually a nun's face appeared to fill the darkened opening. We placed 5 euros in the offered box and it disappeared to be replaced a few moments later by a box of biscuits.

Traffic struggled up the cobbled steep main road from the west with no more than an inch clearance from the buildings on each side to get into the main square – Plaza del Cabildo surrounded by impressive edifices and providing space for about thirty cars to park. It was a clear-cut case for pedestrianisation. The church of Santa Maria stands on the south side and had won the battle with St Peters to be the more prominent. The main altar and side chapel were fine but we preferred St Peters. We visited the Town Hall and the crenulated sandstone castle that appeared impenetrable, as we couldn't find the entrance so the Dukes of Arcos are safe. We peered over the 150-metre drop off the cliff down to the river below and explored the more recent conurbation on the west side where we found a Lidl and could have bought the same biscuits for a tenth of the price, but without God's blessing.

Arcos de la Frontera was an impressive place but without a frontier man in sight (*Frontera apparently refers to the front between Arab and Christian lands*).

After lunch I washed Boris – what a state he was in after being towed in the rain behind Katie. Three British parties arrived. One couple had been at Orgiva and told us they had experienced strong winds and cold – no wonder, they were in a small tent – how lucky we were to have the comforts of Katie.

We were into March and with thoughts on when to return home we felt an urgency to maximise our visits so we decided, having done towns and cities, to head for the hills and in particular the Parque Natural de la Sierra and Grazalema.

At the foot of El Bosque 20 miles due east, the sun shone, the mountains rose in the distance and we enjoyed coffee near the visitor centre before walking up in to the small town. The route was lined with trees heavily laden with both oranges and lemons and soon we came into a square with stereotypical old men sitting on benches, the town hall and church, a Spanish daily life cameo along with the Repsol truck noisily collecting and delivering bright orange butane bottles door to door.

We headed for Grazalema but then turned along the A372 towards Puerta de Montejague on the CA339 towards Zahara de la Sierra. The blue of the reservoir (Embalse de Zahara) came into view followed by the distant hilltop castle of Zahara de la Sierra. We couldn't wait to get there. Gail drove straight up the hill into the village, saw a *'Parking'* sign, climbed farther then plunged down a hill that seemed to drop from the highest point at 1,100 to the lowest at 300 metres in 15 seconds, then rounded a corner where Gail decided enough was enough and parked. The main street had pretty churches, a police station, and a number of restaurants so we chose one in the main square. After lunch we climbed up the steps to the Castillo where great views across the lake and the surrounding countryside showed how important a defensive enclave this was for Moors and Christians alike.

We started back to Grazalema along the CA531 because on the map it was all wiggly and colourful with many hairpins and switchbacks. It soon started winding and climbing steeply with continuing views of the lake right to the summit at Puerto de las Palomas. Keen to take photos, I parked Boris and jumped out only to fall on my arse as the whole area was covered in ice. We turned right on the A372 missing Grazalema again and thereby the chance of getting soaked in the wettest place in Spain, and stopped at a truly beautiful point showing the weirdly shaped mountains, desolate heights and lush green valleys with native pines that the area is so famous for. It reminded us of the Smoky Mountains in eastern USA. It was a wondrous day that surpassed our expectations. Zahara de La Sierra was one of the nicest villages we had been to (despite the panic attack descending the hill) and El Bosque well worth a visit.

On our return to Katie we received a surprise phone call from British friends on a short break near Malaga who wondered whether we could drive the two hours over the mountains for a meal. We arranged to meet up for lunch at their hotel on the Friday.

It was a week since we had done tank duty in Granada so with beautiful sunshine, and clear skies and Gail's washing underway we decided to dump the tanks and fill with water. I reconnoitred the route and reckoned we could make it round to the service block at the back of the site if we watched out for a couple of trees. The 'dump station' was a 6-inch diameter hole in the ground with a plastic tube liner. It had no surround, only grass which had a profusion of growth no doubt owing to the nutrients in the spillage. A metal flag inscribed 'WC Quimique' marked it.

We filled the water from the push-button taps, Gail holding her tap in for the hose and the city fill whilst I used our water bottle and the gravity fill. It took a long time but to make sure we had a full tank I levelled Katie with the jacks but they started to sink into the spongy ground. I crouched to look underneath then stood up and gashed my head on the sharp metal 'WC Quimique' sign, drawing much blood. We spent the afternoon cleaning Katie but Gail was more anxious that I 'don't overdo it,' fearing concussion.

The next day was a blazer. The girl from the tent pitched near us popped over to offer a book exchange, which we welcomed having exhausted our own supply.

We had a good drive via Ronda and the long downhill run to the coast to meet up with our friends at their hotel. They remarked how fantastically healthy we looked and we felt it. We had a great day enjoying drinks around the pool and later a beachside meal, but it was strange as we hadn't heard so many British voices or seen so much white corpulent flesh exposed for six months.

The weather was sufficiently hot by 9.30 a.m. to sit out and sunbathe and the light persisted until 7.30 p.m. in the evenings but we were at the end of the first week in March and we had to think about returning home. Should we go north through Madrid or go east and up the Mediterranean coast? Would the former be safer but then how would we get on driving through Madrid? We decided to move on Wednesday, March 10. The decision came with all types of emotions and questions attached: – our trip was coming to a close, the weather may be worse, should we stay and enjoy it for a bit longer, do we want to look round more cities, are we architecturally satiated, would the toilet chemicals and biodegradable paper last? The weather at home forecasted sleet and snow with night temperatures of minus 2 degrees centigrade.

My left hand had developed eczema over the past week and it had worsened, spreading to my elbow and eyelid, what was causing this? I hadn't had an episode for many years, although it had been a major childhood affliction behind my knees and was only cured by my long-delayed transition from short pants, which wouldn't wear out whatever torture I put them through, to long trousers. I wondered about the camp water. I have made little mention of the campsite itself apart from our dumping activity. It had been quiet, with only two to three arrivals a day and they never stayed longer than a couple of nights so we could sit out undisturbed. It also meant we had little company. The campsite looked pretty in the sunshine with all the mimosa trees a profusion of yellow blossom, a complete contrast to the day we arrived in pouring rain. The electric was intermittent. The staff were friendly and worked hard to maintain the site grounds but seemed to ignore the toilets and showers, many of which were broken or dirty. Access to waste dumping was reasonable but the drain had to be unblocked each time we used it. We guessed facilities would be stretched in season or when motorbike rallies were held in nearby Jerez.

The young guy in the camp office said Internet was available next door to the Carrefour *supermercado*, where I found a modern dedicated facility with about twenty Internet stations and for 1.50 euros I got the forty-five minutes I needed to read the accumulation of important emails. Lacy Levine wondered whether I was finishing in the first five whereas Eliza knew I craved to shoot like a film star with huge explosions. It was unsurprising therefore that Faith was sure I wasn't too happy with my life recently as I had increased weight, less energy, bad moods and wrinkles on my face. Thank God for Jay who could turn me into a superman, in bed, in showers and locker rooms and Marsha who could make me adore her whatever my problems. Peter and Jenny had stayed at a site near Sevilla but left after one night as it was a dump of a place and moved on to Portugal where they had suffered with the rain and had to wade through water to get to their caravan and tow others out with their four-wheel drive.

We decided to move on to Cordoba so paid our 207 euros bill for the fourteen nights stay at Arcos de la Frontera including electric. The guy at reception agreed to phone ahead to the next site to check they had sufficient access for Katie. Isabel said 'there would be no problem' – where had we heard that before? We did another dump and fill in preparation.

Cordoba and the Journey North

By 10:00 a.m. we were ready to depart Camping Lago de Arcos – why not, everybody else had? I drove out into the mist and onto the blind bend on one of those do-or-die missions between leaving it too long to avoid Katie hitting the fence across the road or too short and clouting Boris on the gatepost. We waved our appreciation to the ashen-faced drivers forced to jam their brakes on and some waved back in a Spanish-style salute we didn't recognise.

To get the benefit of motorway driving we added 20 miles and took the A382 west but when we ran over the cobbled sections for Tractor Tom or his Spanish equivalent to cross the road every bottle and can in our cupboards and fridge leapt around like some manic break dancer.

The motorway north to Sevilla was a relief but became busier and bumpier on the outskirts. We made a stop at one of the few places you could for a quick snack and refuel and had no trouble finding the grandly sounding *'Camping Carlos III'* where we were greeted by two pleasant girls in Reception including Isabel. It was a large site with many cabins and mobile homes, which despite their illusions to mobility were definitely there to stay with awnings, windbreaks, wooden patios, tables and chairs, mats, outside cookers, sinks, potted palm trees, bicycles, and swing balls all protected by Ronsealed wooden fences.

Confident after six months driving experience and with a watching crowd I headed for a suitable pitch to find that the sound of a branch rubbing on the roof meant a branch was scraping on the roof and turning would result in something nasty happening. A car towing a caravan, feeding off my confidence, pulled in close behind. With Katie and Boris at an acute angle either side of a tree planted unhelpfully on a corner, the A-frame was tight and didn't want to unhitch. A lot of arm waving and chin rubbing and a damn good tug on the A-frame were required before everyone was extricated back to where they started.

We re-hitched Boris to Katie and headed to the lower part of the campsite and open pitches devoid of trees. A gentleman stood up by his caravan to cheer us on as we passed and raised his hat for emphasis. Having parked, he ran over to get a better look at Katie continually calling to his wife to come over as well. They were Finns, he spoke Spanish and she supposedly English but she never uttered a word, apparently embarrassed by her husband's exhibitionism.

The electric supply was two cable lengths away but after we had blown it twice by only plugging in we decided to use our own resource and switched the inverter on. My eczema was no better and so I went for a twenty-minute hot shower in the new super-clean shower block to try to rid myself of any lingering residues from the Arcos water.

We would have enjoyed driving by Cordoba's fine historic buildings if we had not been lost looking for a car park by the *'Puerta del Puente'* eventually ending up near the town walls on Av. de Corregidor. Afterwards we did a six-hour day touring and felt we had done justice to this capital that started as a Roman port and became the capital of the Moorish Kingdom of El-Andalus and a city where in the eleventh century Jews, Muslims and Christians lived side by side.

The dusty entrance gardens and close proximity of other buildings left us unprepared for the colossus of the Mezquita or Mosque (AD 785) re-dedicated to Christianity in 1236. The guidebook said the Christians were so in awe of the mosque's structure they built their magnificent cathedral inside but it took them 234 years and maybe they were just using it as a roof covering. Stepping into the Mezquita we were stunned by the canopy of red and white arches atop a forest of columns pinched from the Roman remains scrapyard; those columns too short were bolstered up, if too tall they were buried deeper. The mosque was one of the largest in Islam but still didn't face Mecca; today's philosophical excuse was because Caliph Abderramán I expressed in his poetry how much he was missing the mosques of his hometown, he had it point to Damascus. But let's face it anyone can make a mistake and in 785 Mecca was a long way away and maybe sundials weren't what they were in later years.

Reflecting that the Romans were the city's founders; we visited the sixteen-arch Roman Bridge that was functioning as a conveyor of modern chariots into town. Looking across the bridge, somewhat unenthusiastically, we saw the Torre de Calahorra (Arab fortification) and the uninspiring derelict Moorish Olive Mills on the side of the river Guadaquivir. We spent time meandering through the narrow streets of the Barrio de la Juderia or Jewish Quarter and another thirty minutes of disorientated alley work looking for *'Calleja de Flores'* because it looked pretty in the guidebook and on postcards but turned out to be a back alley with a distant view of the cathedral tower and a few limp flowers. The synagogue also proved elusive and unhelpfully had no signage outside. It was a bit squidgy, its claim to fame being that it's the only one to have survived from the fifteenth century. The *'Casa del Indiano'* had found fame because someone had built an upward extension on top of the original Mudejar walls. They had cool patios and Charlie Dimmock water features.

To escape the maze of narrow passages, we followed the streets up to *'Plaza de las Tendillas'* where we hoped to find *'the pulsating heart of Cordoba and meeting point for locals and visitors alike'*. It was pulsating with thousands of students surrounded by police in full riot gear which made me wary. Many years previously I had accidentally got caught up in a student demonstration in Milan and ran round a corner to escape only to find I was facing a street full of tanks coming towards me like some re-enactment of a Red Square scene. We took avoiding action by going into a Farmacia to buy Alka-Seltzer. After lunch, at a restaurant with a mosque-like interior, and thinking we had done almost all of the tour we came by chance to the *'Alcazar de los Reyes Cristianos,'* The Palace of the Christian Kings where Isabel and Ferdinand met with Christopher Columbus and known for its wonderful formal water gardens. Despite an already long day, we couldn't leave without adding it to our tour. Returning to the car park we expected a large bill but the attendant had left for the day and we had no one to pay.

The magnificence of the buildings in the ancient city contrasted sharply with ugly high-rise tenement buildings festooned with clothes drying on the balconies that we saw on the way out.

Once back in Katie, I switched on the TV to horrific and disquieting news of a major terrorist bomb attack at the main railway station in Madrid that had killed 190 people and injured over 1,200 and the reason for the crowds in the square. We saw demonstrations on TV particularly in Bilbao, the heart of the Basque country and where we would be going.

The day after the bombing the grey of the rain-bearing clouds reflected the mood of the Spanish nation. We were in one of those, should we, shouldn't we do something frame of minds. I searched on the satnav for local CEPSA garages and with a shortlist we loaded the propane cylinder in Boris and went on a bottle hunt. Gail drove us south down the motorway to kilometre 441 then cut inland along the country lane of the A386 through lovely rolling hills adorned with olive trees, past the villages of Santaella, Fontanar, left and right to Aguilar, where it seemed impossible to get lost in such a small place but we did and managed to complete a full circle of the narrow streets in a surprisingly busy town. Eventually we escaped and as we approached the N331 we spotted the targeted CEPSA station. A crowd of men stood around on the garage forecourt all in active discussion but, as I pulled up, conversation stopped and I felt we were being watched and scrutinised. No petrol was being served. I asked the attendant whether he had butane or propane and he replied with a curt 'no'. Come on, guys, do you really think an unusual little white car with GB plates is good camouflage for a terrorist man and woman? Although, on second thoughts, I guess asking for an incendiary cylinder of propane was a bit suspicious. I snuck back to the car trying to act normally but that usually ends up in doing a funny walk that makes you even more suspicious or looking stupid. But this time, I had a more basic reason for the funny walk. Gail drove us onto the N331 in the direction of Cordoba but by now I needed a pee, not one of those 'if you see a toilet be good enough to stop' kind more the 'stop anywhere I'm bursting' kind usually followed by 'I can't stop here we're on a narrow single carriage road it's going to be awkward' and then 'for goodness sake it's already more than awkward, just stop' – but Gail won't do it as she has an iron bladder, and will always comply with traffic law. After what seemed an eternity, a lay-by appeared and before the car had stopped moving I was out, zipper already undone and in action. As the pressure subsided I started to glance around. Only then did I realise my hasty initial scan of the protective environment was flawed and I was in full view of the house that I thought was behind the bush and the traffic coming from the other direction but the flow was still strong and it was just a matter of sticking it out, if you see what I mean.

Fernan Nunez CEPSA had no propane either, however, they did seem a little more interested, perhaps CEPSA headquarters had put out an all-points bulletin on us. We arrived in Cordoba, targeting Avenue de Cadiz in the south. We spotted the CEPSA garage from the far side of the dual carriageway and were those grey bottles alongside the shiny ones? We did a U-turn and pulled up in a Starsky and Hutch screeching manoeuvre inasmuch as one can with a little A140 with Gail driving. Surprise, surprise, he had propane. With great care he selected the rusty old cylinder at the back of the rack that had been rejected by all previous customers.

Had CEPSA headquarters been on the phone? 'Be a good time to get rid of those rotten cylinders you've been trying to get us to take back.' Still, who were we to complain?

Medina Azahara, began in 936 and completed 25 years later as the *'Versailles of Caliphal Art'* was only 10 kilometres from Cordoba so with no other thoughts in our heads apart from admiring the artwork of the rust patches on our propane cylinder we thought we'd give it a go. We negotiated the streets of Cordoba, found our way out on the A431 when it started raining heavily and turned off onto the CP119 signposted to *'Medina Azahara.'* Maybe new excavations were underway. The road was narrow and every two minutes an articulated soil carrying truck came towards us. Having just left a construction site, it was covered in filth that sprayed in all directions, the tyres giving off great flying gobs of mud the trajectory of which increased as the truck picked up speed, dropping it all over the road and any little white cars coming in the opposite direction. Great streams of brown mud ran down the windscreen except at the limits of the wipers where a curtain effect rapidly built up, framing the window.

Ascending a hill we saw this amazing building on the hillside and thought it must be Santa Maria de Trassiera but we turned off to Medina Azahara and parked in a remarkably empty car park. On the entrance door was a small handwritten note – *'as a mark of respect for those killed yesterday in Madrid we are closed'*.

'Bloody marvellous, they're closed because of the bombing,' I shouted to Gail

'That's a pity, we're only here today,' she responded.

'The differing cultural responses of the Americans and Spanish is interesting,' I said back at the car.

'Why?'

'Well, after 9/11 the Americans all returned to work because they said it showed strength.'

'That's true, but the Spanish probably have a more religious approach,' she pointed out.

'Maybe they're marking their respect assembling on CEPSA forecourts?' I responded.

'That's very unsympathetic,' Gail chided.

We could see the ruins below and they had that come and see us look.

We intended to return to the campsite via Posadas but were distracted by a great castle on the hill above Almodovar. We turned immediately into the village and up the narrow street only to find in a one-street village, we had gone back on ourselves. Gail wound down the window to ask an old boy – 'Castillo?' – hand signals said we should turn around again then 'a la izquierda' (left) then more elaborate gesticulations indicating up and up. What a narrow track it turned out to be and then there we were at the castle gate reading more notices. Being a non-governmental organisation they weren't marking their respect by not working but they had just closed for lunch (2.30-4.00 p.m.). We walked around the walls; admired the view down to the river at the bottom and over the never-ending plains beyond and reflected on how in the days of being atop an easily defended hill and able to view your enemy's approach from afar it was well placed. We descended and drove home across the plain from which the castle could be seen for miles.

La Carlota, a couple of miles down the road, has a delightfully long Rambla, and a beautiful church exterior. We wanted to get Boris cleaned up after his mud bath and visited the Repsol garage car wash that had a jet spray but no brushes. We were being watched intently by two gentlemen in city suits and a third less formally dressed. Then we made a major discovery – a car carpet-cleaning machine. It was badly needed as our car vacuum just redistributed the dirt but did make track marks on it so you could pretend it had worked. We posted our carpet into the orifice and pressed every button but it emerged no different from before. The less formally dressed gentleman, who we deduced to be the garage owner, came over to demonstrate. You could pop the mat in and ram it up and down, wet or dry with ample time – quite brilliant. I put the sopping wet mats back in Boris, cleaning enthusiasm on the wet cycle having been unbridled.

We decided to move on the next day and I called the campsite we had selected at La Carolina – *'no problema.'*

I noted the batteries were holding out at 12.6v. And of course we had the new propane.

It rained hard in the night and despite all the gravel many of the lower pitches had large puddles. The sky remained dark with only the slightest hint of a break – should we stay or go? I emptied the water from the awning then started to leather Katie off. The *Dri-Wash* polish seemed to rise to the surface and a wipe with a dry cloth soon restored the shine. Once finished, we decided to make the move. Gail paid the bill of 49.62 euros for the 3 nights.

We were soon hooked up and crept cautiously out of the site but as we approached the gate the owner ran out from the office to proudly present us with a cigarette lighter and key ring with a camp motif. We still have them in the glove compartment, not because we treasure them, but they became one of those things that gets put away and never emerges again.

The sky alternated from dark threatening clouds to fluffy white ones so we presumed the storm was moving north.

We didn't stop and had a good drive up and down the hills that were completely covered by patterned olive groves. As we approached our turn (kilometre 257) from the *autovia* to Santa Elena just north of La Carolina but short of the Despenaperros defile we noticed a large yellow sign '*Exit kilometre 259 for Camping.*' So we ignored it and followed the *Caravan Club* book directions and left at kilometre 257 that dumped us at the end of a lonely exit ramp with no further signs so with visions of getting stuck in the village we blindly went for it. On the other side of the village we found the signs that we would have seen if we had taken the signed route. We turned right up a sharp hill – I was certain we would ground the rear of Katie and did with that awful teeth-gritting scraping sound. At the top of the hill on the left was a stone gateway with two arches that were quite narrow and at an angle away from us with recycling waste bins opposite. With no room for a swing out to make the turn it looked like we were stuck.

A moustached balding man in brown sweater and corduroy pants came running out from the office, smiling and arms waving – '*No problema!*' After more arm waving indicating where we should drive, he set off back into the campground in his little white van that all workers have. So, whilst he was on the inside looking at us, we were on the outside driving alongside the campground fence in a field of olive groves and not looking at him. At the end of the track, and I use that in the loosest sense, he opened another gate and I followed his instructions on the swing in. I must learn not to do that. We were stuck with Katie partially round and Boris at an acute angle behind and impossible to detach. I decided to try something else. I unhooked Boris's brake cable, started his engine so the power steering worked and opened his window so Gail could steer him if necessary. Then I reversed Katie a little way to get a new approach and we were round.

Gail climbed in the white van and I followed. As he had set off with his accelerator foot firmly on the floor he soon disappeared. Fortunately, Gail had the walkie-talkie so was able to talk me in. The site was pretty with the pitches set in a pine forest and surprisingly, fully serviced with electric, water, waste, TV and telephone connections. We positioned Katie then decided she would be better off the other way round. All this manoeuvring played havoc with his lightly tarmacked roadway.

We registered with Federico who spoke no English but communication seemed to flow with much laughter and with brochures he indicated there was a lot to see in the area. The site was sparsely populated but a few more campers arrived later.

The sun was shining, fluffy white clouds floated benevolently across the sky, the TV worked and we could relax. We walked round the village; it was quiet possibly owing to the events in Madrid. Some children on their bicycles shadowed us round the village square and tried their English – shouting 'Hello' and 'Goodbye' but when Gail approached with questions they clammed up just like I did if someone asked me a question in Spanish.

Later the electric tripped out – 'is it really 10 amps?'

Sunday brought the usual phone problems – Gail's parents called us on the mobile – we couldn't seem to get the message across – so we didn't answer and switched the phone off. Gail went to return the call from the phone box only to find our card wouldn't work.

'We'll have to go into the village to find another box.'

'I'll go and chat to Federico.'

In the office I waved our card and pointed at the phone box.

'*No es bueno.*'

He pointed to the phone/fax machine in the office.

'*No problema.*'

I fetched Gail, and then after she had finished I had to be cheeky and ask to use it to call my mother.

'*No problema.*'

And he refused all payment.

It was a sunny start but as the morning progressed it chilled over – where did that sun go? We decided on a trip across country to Úbeda, recommended in both our red and green guidebooks, and inadvertently went via the local Santa Elena Visitors Centre – Puerta de Andalucia – Parque Natural de Despenaperros – gateway to Andalucia – ideal for those visiting Andalucia but less so those leaving. We spent the time doing the 'been there, done that' and ticking off the boxes.

We went down the motorway to La Carolina, noted a Lidl for potential shopping later, and in so doing missed the first turn but took the second one for distracted motorists onto the A301. The road crossed the hills and reservoirs. The whole landscape was a tapestry of ordered rows of olive trees, right over the tops of the hills and encircling Úbeda. During the conquest of Úbeda by the King Ferdinand III in the thirteenth century one of his captains disappeared and missed the battle. When it was finished he came to the city and explained he was lost 'in the hills of Úbeda'. It was considered as a cowardly affair and passed into Spanish folklore something along the lines off: 'Don't go to the hills of Úbeda.'

Entering Úbeda we followed *'Parking'* signs right into the heart of the town and the triangular Plaza de Andalucia where the streets all meet and is the social centre of the town. Ancient buildings including the thirteenth-century clock tower and the nineteenth-century Royal Abattoir surround it. The road slopes quite steeply to accommodate a recently constructed underground car park that seems to have escaped guidebook inclusion.

The congregation was spilling out of the church (Iglesia de La Trinidad) so we followed on the basis that they should be going to cafés – but there were none to be seen so we started our tour. Úbeda is a Renaissance city in the geographical centre of the province of Jaén and designated a World Heritage by UNESCO in 2003 with some fine buildings but they had a lot of graffiti scrawled on them. The prettiest spot was Plaza de Varquez de Molina in the old part with neatly clipped hedges and a row of orange trees and surrounded by some of the finest buildings including the imposing thirteenth-century church Santa Maria de los Reales (Santa Maria of the Royal Citadel), the Sacred Chapel (Sacra Capilla del Salvador) with the Parador hotel on the left and old command gallery on the right along with the town hall and police station. Next to the church of Santa Maria de los Reales Alcazares is the Carcel del Obisco or Bishop's prison – the story has it that this is where misbehaving nuns were sent for punishment, by the bishop, no doubt. It is now the town's courthouse. The buildings were Renaissance-style, solid and in sandstone with attractive features like loggias high up and many lions to commemorate the knights of Úbeda who emulated that beast when fighting against the Moors.

We wandered into some narrow streets and pressed ourselves against the walls of buildings as a cars squeezed past. An old lady in black with a white apron laughed from across the street. Ignoring her, we crossed over to look at a tourist information post – but she wouldn't have it and came up gesticulating that we should follow her inside the building behind. Gail is charming to everybody so we followed into a pretty courtyard-patio surrounded by an upper wooden veranda making approving *'bonito'* noises. She took Gail's arm, led her into the building and then drew back a blanket-like curtain, and into a cave that was lined with bric-a-brac and ceramics for which Úbeda apparently has a number of famous artisans (well, famous in Úbeda) and also seemed to be her bedroom. I took a photo of her with Gail and quick as the flash on the camera she had her hand out for money – I gave her 2 euros and mentally calculated how many she had to do each day to get a decent standard of living over and above her pension and custodial duties. What a sales entrepreneur, shame she didn't serve coffee then she could have really made some money. We didn't find a café but did find a *chocolatier* (or should that be *chocolatero* or, was it pretending to be French?), one of those with tray after tray of beautifully designed and arranged chocolates often involving strawberries, even the church across the street had been modelled grand-scale in chocolate along with a chocolate cake showing a football pitch and players and *'feliz cumpleaños'* – happy birthday iced on it in a lurid blue.

You gaze through the window, saliva drooling, knowing that if you chose something you'd still come out of the shop and look back in the window wishing you'd bought something else. We managed to pass through the chocolaty aromas to a *patisserie* section at the rear of the shop where we consumed a coffee and custard.

We returned to the campsite and as we entered our friend Federico ran out from the restaurant waving his arms like windmills– what has happened – has Katie burnt down; what tragedy awaits us?

'*Ma famillia – auto caravana*'

'What does he mean have his family wrecked our motorhome or do they want to buy it?'

'No I think he wants them to visit.'

'*Diez minutos*' I shout back, '*No problema.*'

By the time we had parked and exited Boris, thirty people had managed to cover the 300 yards from the family lunch in the restaurant to Katie and were ahead of us waiting at the door. Grandmas, grandpas in tartan slippers, children of all ages and dimensions, loving couples, people in their Sunday-best anoraks, sweaters and skirts or, as gender permitted, sports jackets and creased trousers all shouting to each other (normal talking in Spain) and jostling for a front-row position.

'Stand back I have the key, *tengo la llave,*' I shouted above the crowd's noise.

Reluctantly, the crowd parted. I took off my shoe in a demonstration of good housekeeping waved it over my head and shouted:

'*Zapatos,*' not knowing the word for remove.

They laughed; they probably thought we were Muslims.

Gail attempted to show them round inside but that was difficult to do once more than fifteen had entered and were in Harrods's sale mode. Every drawer and cupboard was opened, they sat in the toilet, put their heads in the shower, stood at the cooker, opened knicker drawers, sat at the dining table and pushed people aside so they could get the fridge open.

'*El dormitorio es enorme,*'

'El refrigerador es enorme,'
'La microonda es enorme,'
'Es bueno,'
'No, es fantástico,'
'Un palacio,'
'Si, un palacio,' they all shouted to those still trying to enter.

The press of females getting in to Katie outdid the men or, perhaps it was not the man thing to look at domestic arrangements, although it could have been the shoe removal that spurred their interest in the outside mechanics of Katie. I had to open all the lockers to show them the generator, plumbing, electrics and the engine or as much of it as you can see. Each man at the front of the throng took my words and turned to interpret to the rest of the men. As I was saying almost all of it in Anglo-Spanish with sign language goodness knows what was being translated. Word must have reached the village because the local breakdown truck arrived with a dozen men hanging on the back of it and I had to do the mechanics tour again. Once all had been round and put their shoes back on a group photo in front of Katie was requested to hang in the village hall. A chubby girl of about ten in a pink velvet tracksuit who had made sure she was to the front of the photograph was anxious to give me her name and address so that I send her the photo. I promised to email it to Federico.

Later a few other motorhomes arrived on site; many were Brits returning home like us. Halfway through March and we were still away. Despite a bright, sunny start the weather forecast suggested cloud was over the southeast corner of Spain and we were soon covered by it. We decided a mountain tour would be a waste of time and went to Lidl in La Carolina to stock up. On return our new neighbours told us of a great little bar where they spent the afternoon having drinks and *tapas* and were going for an evening meal – we said we might join them.

We decided to try for a move the next day so I spent most of the afternoon phoning to try and find somewhere north of Madrid that would accommodate Katie. A warmly spoken girl at Gargantilla del Lozoya said:

'La entrada es aceptable.'

'La parcela no problema.'

'Los árboles un problema posiblemente – tiene una parcela especial para usted – no problema.'

So, the trees might be a problem but they have a special pitch for us, no problem.

We dumped the tanks, which was so convenient with the drain on the pitch and with water available at high pressure we gave the holding tank a thorough rinse. We stuck Boris round the back of Katie with the A-frame on for a quick getaway. We paid the 56.65 euros bill and spent the evening in a little bar in the village with Georgie and Pete, who had an American pick-up truck, a small tent and a Harley Davidson. They were renting a flat near Fuengirola. They and Gail enjoyed pre-dinner drinks of coffees laced with brandy accompanied by *tapas*. We discussed how well Pete and Georgie got on with Spanish and they told us of a friend who ordered *'sopa de mariscos'* (seafood soup) but actually asked for *'sopa de maricones'* (homosexual soup). The lady owner laid a table especially for us with the best paper tablecloth and soon it was groaning under the weight of food and wine. We were the only diners in a small village bar watched over by old men who had heard there was live entertainment that evening. We left our friends and the old men to more drinking while we headed for a supposedly early night prior to departure the next day.

We rolled out early and did without our usual showers ready to put some miles on and tackle Madrid's traffic. Everything was geared towards leaving. Boris was already hooked up, but since it had rained in the night I had to remove water from the slide awning and clean the windscreen. Others were also up and about doing their tasks; it was like an old-fashioned Formula 1 Grand Prix; who would be through the front gates first? Well, not us, because we didn't fit through the gates so I had to go and get the key from Federico and we manoeuvred our way through the rear gate and along the field to the front entrance where Federico was waiting to say goodbye, present a miniature bottle of local olive oil to Gail and for him and me to have a photo taken, arms around each other's shoulders. At the bottom of the hill we proceeded slowly but still produced that nerve-jangling grinding from the rear – but what could we do?

Despenaperros translates to falling dogs in English, a reference to the sheer craggy cliffs of the precipitate gorge in the hills that mark the border between Castilla La Mancha and Andalucia. The road E5/NIV dips, weaves and climbs through the gorge, the southbound road taking a completely different route to the northbound. We drove through torrential rain, Katie leaving a cloud of spray and when I looked in the rear-view camera Boris had completely disappeared. Some of the roads immediately after we left were pretty rough and didn't improve much as we headed north to Madrid. Over rough roads Katie had an annoying and distracting squeak somewhere in the cabin that we had been unable to track down. It was just what I needed before driving through Madrid but with the aid of the satnav all went well and we took our second break of the day at a truck stop before setting off again down a country road towards the campsite.

Something important happened here – I was speculating more and more about the *'tweetie pie'* squeak that appeared to be caught in Katie's dashboard and wondering whether I would have to remove it or the engine cover to investigate. In a moment of inspiration or desperation I asked Gail to check the two rubber tent pegs she had stuck between the dashboard and front window to support the *'TOWING'* sign and the squeak that started from Fuente de Piedra 591 miles previously, stopped.

Snow was visible on the hills as we drove under a viaduct, turned sharp right and then left up the hill to the log cabin reception of *'Camping Monte Holiday.'*

The cheery campsite manager guided us to a large open grassy spot protected by trees, we had lunch then set Katie up – once more blowing the fuses on three outlets so the guy had to come over again. The weather became sunny and warm and we sat out, relaxed, swatted flies and read the guidebooks surrounded by snow-covered Sierra de Guadarrama Mountains; we could have been in Switzerland or Austria.

The following day keen to explore the area, we drove along the M604 towards Rascafria. As we left and climbed, the scenery became staggeringly beautiful, the altitude markers started to fly by and we were soon at the 1,840-metre summit, Puerto de los Cotos, where the roadside snow was deep with red and white depth markers poking through. As we cleared the peak we were enveloped in a dense mist. The narrow road became difficult to see and the situation frightening for Gail driving. Yet, it was more dangerous to stop, as approaching cars would suddenly appear out of the mist, their headlights diffused into golden orbs. We descended the steep hairpin bends relieved when the mist became less intense and like wispy steam from a cooling kettle finally disappeared. We had dropped into a lush wooded valley and entered San Ildefonso La Granja – described as a *'Mini Versailles'*. We had a welcome, but surprisingly expensive coffee and pastry then wandered up the central avenue towards the Palacio.

Entry to the Palacio was complicated. At the first attempt we were flatly denied access even though they issued free tickets for members of European Union on a Wednesday. Then we were told to come back at 12.30 p.m. which we did. We put our stuff through the X-ray scanner and were, rather hastily I thought, ushered through a locked door into the palace a great stone staircase ahead of us. Alone, we wandered freely round the vestibule, in the way we like admiring the exhibits, but at the same time following the red ropes, then ascended the stairs crossed through a couple of magnificent rooms where by chance we caught up with a tour group and Spanish guide. Knowing how wordy some of the guides can be and wishing to continue the tour at our own pace I squeezed past the group to be manhandled by a guard in a uniform more suited to an admiral of the fleet who said:

'Grupo?'

'No, solo,' I answered.

'Grupo solamente,' he said.

He must have been mightily impressed by my command of Spanish thus far but how could I explain that a two-hour tour in Spanish was going to be more taxing, thoroughly boring and did he really need a gun in a big white leather holster in a museum?

'*Solo, solamente,*' I tried.

'*El grupo o la salida,*' he said rather forcefully.

It's only a bloody museum, I thought, what is all the fuss about?

'*La salida,*' I said and was immediately frogmarched through the room (exquisite furniture and paintings) down the stone stairs (elegant suit of armour and balustrades) and out of the large wooden doors (magnificent carvings) into the courtyard, Gail running behind to catch up.

'They seem a bit sensitive after the bombings,' I said, taken aback.

'And I don't think we had our money's worth.'

'We didn't pay any money.'

'Well I mean the EU didn't get full value.'

With unexpected time to spare, we headed for Segovia. According to the guidebook we expected a great view from the east but were disappointed. We parked in the helpfully named '*Aqueduct Car Park*' close to the old town. Our first view was an early twelfth-century church – San Millan with Romanesque features (there are more Romanesque churches in Segovia than anywhere else in Europe). Then as we proceeded along the road we could see the cathedral off to our left, some half-timbered houses and at the end the Aqueduct built in the first century by the Romans, and still working up to the nineteenth century. 728 m long, 29 m high and 118 pillars so well fashioned out of stone blocks from the mountains we had driven over that morning that no cement holds them together. 2000 years of jaw-dropping history spanning the entrance road.

As we climbed the twisting narrow streets that led away from the aqueduct, we were fascinated by a basket in a restaurant window containing three cute smiling porcelain pigs and debated whether that really encouraged you to go in and eat? Then we entered the Plaza de San Martin with its statue of Juan Bravo, the Iglesia de San Martin. We walked alongside the church to the cathedral then away again down narrow cobbled streets towards the Alcazar (Castle). We were struck by the magnificent views everywhere – back to the cathedral, down from the Alcazar to the monasteria and church, and the front entrance of the Alcazar. The Alcazar is everyone's dream of a castle with conical turrets and tower and will be familiar even if you haven't been there as I learnt later it was used as the model for the Disneyland castle. It has also had a make-over as the original was burnt down in 1862. Nevertheless, as I'm sure people do in Disneyland, our imaginations run riot and inside we enjoyed the Galley Chamber and Hall of Monarchs. It was in this castle that Queen Isabel promised Columbus the financial backing he needed to discover America. It had knights in armour on foot and on horseback and is that St George of England we see in a tapestry? We climbed the high tower and overtook wheezing children and adults to take in wonderful views of Segovia's townscape, the cathedral, the Alcazar looking west, the tower wall and the countryside where I imagined Don Quixote riding across the hills tilting at windmills.

We drove back along the main road N110 to the N1 E5. It was a great day but a shame the mist hadn't lifted.

The village of Pedraza, accessible using the tiny SG612 that crosses the Sierra de Guadarrama, was selected for our final day's tour. The sun shone and the mountains looked clearer. We drove into Lozoya before climbing up a narrow road into the pine forests before emerging to marvel at views of the snow-covered Sierra Guadarrama. Soon we were into snow ourselves and through the Puerto de Navarria at 1778 metres by a rescue hut. As we descended the vista of the walled city of Pedraza, its church and castle on a commanding hill opened up to us. We drove in through the arched gateway set in the old stone fortifications and followed the many signs to park near the castle. Even on the drive through the village we were entranced. We walked back along La Florida to the Iglesia de San Juan and entered Plaza Mayor, an amazing huge open space surrounded by double-storeyed, medieval buildings with verandas. It was a time warp and a Marie Celeste situation – no one was there.

We tried to find a coffee shop. Although tables were set out in the square, nobody could be seen in the restaurant. I peered into the small darkened window of another and jumped as I realised an old man's face was staring back. He unlocked the twenty-five medieval bolts on the door and eventually brought us two of the worst coffees we ever had on our whole trip. They seemed to have been made from last night's grounds and he must have been waiting for the gas lorry to deliver, as they were cold. Holding the coffees in our cupped hands to warm them up, we could identify the Palace de Los Marquesas de la Finesta on the left, the Calle Real Palacio then the Town Hall with its flags flying at the far end, with a restaurant between. The building with the café functions as four houses. One old codger we realised was on a bench at the far end, our sole company – it was hard to determine whether he was alive.

From Plaza Mayor we set off down Calle Real Palacio just to the right of the café. Unexpectedly sighting a *Tabacos* and needing to buy stamps, we were flummoxed to find it closed but then two men came along, rang a bell on the shop next door and strangely the *Tabacos* opened. We waited patiently, only to be told we had found the only *Tabacos* in Spain that had no stamps. Later we undertook a treasure hunt to find the Post Office, clearly marked on the map, but hidden behind a tiny door and then a hole in a wall. It was closed. Going down the Calle Real eventually brought us back to the gateway and the outside walls. The streets were eerily quiet. We walked back up through the village, marvelling at the restoration on Communidad de Villa y Tierra, to the impressive Castillo with particularly fine projecting barbs on the door and views of the land beyond. But like everything else it was closed. Pedraza was a great discovery and worth seeing, if they could add some people it would be perfect. We drove back along the motorway.

Having decided to move the next day, I phoned a campsite that couldn't fit us on a pitch but could offer space in the car park outside. It sounded fine for one night. We turned Katie round and prepared Boris. Our bill came to 55.89 euros for three nights.

On Friday March 19 we were on our way by 8.45 a.m.; some sort of record. The sun was already shining and for the first time the hills around us were clear and the snow line easily visible. We almost felt hard done by at what looked like the beginning of a beautiful mist-free day and a chance to see the area at its best. We drove down the hill to the bottom of the road where we turned left away from Lozoya and the mountains to the west.

We soon joined the *autovia* but the road was unusually busy and we couldn't think why. As we progressed north the weather changed and soon we were in thick fog and mist obscuring anything beyond the road. With so many cars we received the usual curious stares as folks drove by. As we drove towards San Sebastian on the A15 the road cut its way along the Pyrenees through gorges and tunnels and was quite magnificent. We tried a phone call to the campsite but were unable to make ourselves understood. Kilometre 123 did not indicate Lekunberri but kilometre 124 did so we took it and with some comfort it also showed *'Campings.'* We descended round a sharp hairpin bend and spotted the campground where a reception committee was already outside ready to greet us. Neri, the girl I had spoken to the day before, directed us into the car park alongside *'Camping Aralar,'* it had taken us 5 hours 10 minutes despite the heavy traffic that we later found out was because it was Fathers' Day. The campsite itself is pretty and the setting similar to a Swiss mountain village. We repositioned Katie then had coffees in the bar. Gail explored the town while I consulted the maps to plan our next stop. Looking out of the window I saw a large brown rat run down the path beside Katie and I decided not to mention it to Gail on her return.

The camp provided a full laundry service; the TV worked just fine and on reflection the weather after the early mist had been perfect. The next day we would descend to the coast and France but where would we stay?

France and the Journey Home

Saturday's initial objective was the A13 Bidart service station beyond St-Jean-de-Luz in France, about 56 miles away. The drive across the Pyrenees was spectacular with tunnels, six-percent drops and inclines, but too intensive to take in the scenery. I was however conscious of an increasingly odiferous, earthy stink in the cab. After we had both accused each other of foul deeds Gail realised that all the swishing around had induced the broccoli water from the previous night's dinner to vaporise through the trap in the grey tank to permeate the cab with an offensive stench. We joined the A63/E80 and traffic increased. We stopped for probably the last fuel in Spain but had to buy '98' as the '95' pumps had a height restriction. As we approached the French border I reflected on the fact that despite all the warnings and our own trepidations we had encountered no endangerment in Spain, on the contrary, we had felt completely safe.

The usual quandary arose about which lane to go in, should it be trucks, or cars and buses? We went for cars and buses but it converged into a single lane with a 3.5 metres height restriction so I made a fast turn between the bollards into the truck lane. Out of the corner of my eye I saw two uniformed guards starting to sprint for the booth but it was too late I accelerated Katie and we were through before they could stop us.

We pulled into Bidart services and found an easily accessible motorhome service point. Dumping the tanks allowed us to get rid of the vile broccoli smell that had plagued us all morning but did nothing for the perfect family of Mum and Dad, boy and girl lunching at the nearby picnic table. After apologies, we moved to the truck park to have our own lunch, and get away from the smell of our discharge. We started thinking about where to go. We had read good reports of St-Jean-de-Luz in a *Sunday Times* article and *The Rough Guide*. The phone box only accepted cards but not our Spanish one so I called a number of sites on my mobile but they either had answer phones or, didn't answer; it was 1.00 p.m. after all. The one that did answer said Katie was too big for them and they recommend Urrugne, south of St-Jean as suitable. It looked complicated to get to so we decided to keep driving for another 2 hours and then look for a site even though Gail confessed to a headache.

We pushed on up the A63 using Katie's French satnav disk and she seemed much keener to tell us about the turns we shouldn't take than on the Spanish disk. The roads were straight and flat for miles, the *péages* more frequent and the toll money flowing fast. As we drove along I noticed French drivers were taking a keener interest in us than usual and Gail asked a couple of times if a police car was following us. Around 3.00 p.m. we pulled into a Service Area prior to junction 20 and the start of the next section of motorway.

'Are you sure there isn't a police car behind? I can hear a siren,' Gail asked again.

'No, I can't hear anything – anyway he's not going to follow us into a service area with sirens blazing.'

We phoned *Camping Municipal Lou Broustaricq* at *Sanguinet* who spoke English and could take us. As we pulled off the main road, Gail was convinced she could still hear something.

'Do you think Boris's alarm is going?'

'Why would it do that, he's locked up no one can be breaking in at the speed we're doing.'

As we turned off onto a small quiet road I had to admit I could also hear the noise so I stopped blocking one side of the road. I descended from the cab.

'You were right; it is Boris's alarm. I'll just zap it with the remote.'

'I can still hear it.'

'I know it hasn't worked, so I'll start the engine.'

'Won't it start?'

'No, the battery is flat.'

'The alarm is still going so how can the battery be flat?'

'I don't know it's getting on my nerves now.'

'It's not doing my headache any good either.'

'I'll have to disconnect the battery – that'll stop it.'

I struggled removing Boris's battery cover from the driver's floor area and one of the cables to the terminals. Boris's alarm continued but now sounded like a wounded cat 'meep, meep.'

'How can it do that if you've disconnected the battery?'

'Dunno, must have its own battery to outwit idiots that try to disconnect the main one. I'll try to start the engine and hope everything resets to normal.'

I started Katie's generator to provide electrical power and ran an extension cable from one of Katie's locker outlets to Boris. I connected the battery charger and had enough charge in to start Boris. This cancelled the alarm.

We pushed on along more perfectly straight Fen-like roads through pine forests and into the village where, at a roundabout, multiple campground signs appeared. Although the satnav said 'turn right,' we couldn't see a sign for our site, so we ignored the instruction and ended up on the narrowest of roads by the lake. The satnav auto-corrected and we followed the instructions blindly, ending up at the same roundabout and the road it had wanted to take us down originally.

After a couple more turns, we were into the campground but emitting a continuous car alarm that I could do nothing about. Whereas Katie and Boris normally commanded attention, on this occasion we demanded it. Gail went to the office to book in and the girl was on the phone to someone who wanted to know every last nauseating detail about the campsite and its suitability. I sat outside trying to ignore the noise emanating from Boris and the crowd it had attracted, all no doubt wondering why I didn't switch it off or, was it a new UK warning signal? *'A-frame towed car coming!'*

Eventually we drove onto a site in amongst the pine trees but Boris's alarm continued bleeping until I was able to start the engine. We were tired after only 152 miles driving but Boris hadn't helped. What had caused the problem? When we first set off from the UK I had always left Boris unlocked with his keys in the ignition so as to have the steering free. Subsequently, and I couldn't recall where, perhaps Almerimar, I had played with the keys and found I could keep the them in the ignition, lock three doors manually then lock the driver's door with the second set of keys without activating the alarm. So what had changed? It was a mystery.

The site had electricity and waste disposal on each pitch and many of the pine trees a camper told me had been removed because one had blown down and killed a girl. The pitches were spacious and many would have been able to accommodate Katie. The site roads were good and the trees unproblematic. It had that quiet end-of-season feel with a large number of fixed bungalows and chalets. As far as I could determine, the washrooms were unisex and could have been better in the cleanliness department.

I had a bad night suffering from what I can only describe as 'wind' in the chest region. I had suffered this before – was it the strong Spanish coffee? And whilst on medical matters the eczema was no better spreading from my left hand palm and finger, to the right inside elbow, left eyelid and also left neck. I stopped using suntan lotion; as well as all other lotions and potions. I still thought it was due to the water we took on at Arcos de la Frontera.

It rained in the night, ceasing mid-morning. After coffee (yes I know), we decided to take up the recommendation of a Brit I had spoken to at length the day before and walk to the lake. We left the rear of the campsite into some woods and after a short time at the huge and scenic lake (*étang*), then diverted to Sanguinet itself where, surprisingly, we purchased Saturday's *Daily Mail*. Farther round the lake were the boat club and a lakeside restaurant. Here, along with a French couple and their family of three daughters and a baby, we ate heartily. Gail enjoyed a farm salad including duck and Bayonne hams whilst I had the 'Speciality Galette' with hams, *lardons*, *crème fraiche*, *machons* (?), tomatoes, and cheeses with an egg on top and all for 26 euros including wine and coffee (I know) and a 2 euros coke. We were pleased to linger over the food to let another bout of rain pass – all part of recovering – before we set off again. We planned to withdraw the next day but what of Boris? The battery had been charged to the maximum and I drove him around the site without a bleep. Only when he was being towed would we really know.

The next morning after a rainy night we hooked up without any problems and scanned the skies as dark clouds threatened to invade the blue bits. But it didn't matter; we were now definitely in a journeying home mood and we just had to put in the miles.

Our route was to go back under Katie's good guidance into Sanguinet village, and then to the D216, a long, straight, single-lane road that was wet from the night before. After Mios we joined the motorway leading to the A63 and were soon going round the east side of Bordeaux and then north. We stopped for fuel at a BP station negotiating a narrow exit and again farther on at an *aire de service* for lunch. Parking alongside the trucks we felt more at home. The weather alternated between horrendous rain and bright fluffy clouds. At times the road was flooded with surface water and was almost impossible to see the road markings. Then as the sun shone a mini-rainbow was created in the spray of each passing car.

Prior to Poitiers we made another stop and checked our routing to the camp at St-Georges-les-Baillargeaux, who confirmed they could take us without a problem. We followed signs to Futuroscope then St-Georges but Katie appeared to have it organized, picking up the N10 for a few hundred metres then dropping off again. The entrance was wide, parking no problem, the lady welcoming. She directed us to park on the tarmacked road alongside a large pitch, which they were re-seeding. This allowed us to keep Boris on the back. I connected the electric after resetting the fuses in the box, jacked Katie up and was going for the gas cylinder when hailstones started to fall, I jumped into Boris, Gail into Katie and we sat it out for fifteen minutes marooned in our vehicles. I tried to start Boris but the battery was flat – 'Damn!'

Eventually I put Boris on charge, Gail did the washing, as there were clean facilities with dryers and then we took advantage of the hot showers. For once we felt cold.

The next morning was even colder. We paid the bill of 15.40 euros, I gave Boris's battery another boost and we were away by 08.45 a.m. equal to the record. Katie's satnav hadn't kicked in by the time we went through the exit gate so I took an impromptu decision to turn right but it was a mistake. Katie's satnav was all for making a U-turn but that was impossible. We were lost within 200 yards of leaving the campsite. Katie recalculated and we placed ourselves completely in her hands as she took us round the back streets and country lanes, through small villages, over 3.5-tonnes, weight-restricted bridges, past a pretty chateau that Gail missed because her eyes were closed and eventually right back to where we should have been – 'Well done, Katie!'

On the motorway we relaxed but the rain hadn't finished with us and we had more downpours along the way. We made a fuel stop and, fingers crossed, Boris was behaving himself. Later we made a stop at an *aire de service* and looked at directions to the next site at Rambouillet but couldn't find the street name on the satnav. The *Caravan Club* book had – *'by a lake, 2 kilometres east of Rambouillet'* and the *'D956'* were mentioned. I could see a picture of a lake and on this basis I scrolled through the satnav screen map and found a likely candidate and way-marked it. I will not recount in great detail the tour we undertook as a result of there being two lakes being in the vicinity of Rambouillet or us avoiding signs to *'Centre Ville'* but we hadn't expected to drive an extra 24 miles along so many country roads with bicycle lanes twice the size of those designated for Katie, through such pretty villages, wave to so many astounded villagers hugging mid-road bollards, or be only 16 miles from the Arc de Triumph in Paris. The N188 turned out to be our salvation, a dual carriageway heading south to the A10 and a relief stop at Janvry services.

With some telephone assistance, a calming down period, but little confidence, we set off to try again up the N10 heading north, into town via the second exit *'Rambouillet Centre'* and to the first roundabout, now turn right (*'Camping'* sign) under the motorway, round a couple of streets and back on the N10 heading south for 100 metres then the first exit off. After more signs, a small roundabout and a low bridge back under the motorway we eventually reached a bumpy lakeside road leading to the large entrance of *'Camping de l'Etang d'Or,'* which after a 51-mile circular detour looked good.

We benefited from a warm welcome; we wanted to tell them about our massive detour to get more sympathy but couldn't be bothered with the translation. The site was set in a wood at the edge of a lake, had good-sized pitches, with electric, water and waste on many and a good motorhome dump station. It was quiet with only a few occupants. We collapsed and I gained relief therapy by writing my diary for the day.

It was cold in the night; we definitely felt we had moved north and to recover from yesterday's debacle, we decided to stay an extra day. The vehicles were dirty; Boris looked like someone had thrown a bucket of wet sand over him, and as I hosed Katie down, great tides of filth ran off her. It took the efforts of both of us to get them presentable.

We walked into Rambouillet in search of an Internet café walking boldly into a shop that had Internet plastered all over the window but it wasn't and the manager directed us to the *'Cyber Café'* across the park. It was full of students and reminded me of *The Troubadour Café* on the Brompton Road in London during the late 1960s. Ignoring their looks that queried what we were doing there, we went up to the bar and were soon set up at the computer with two coffees. The Internet produced no worries.

Returning through the park Gail posed, Princess Diana Taj Mahal style, on a bench in front of the chateau but we were no longer in tourist mode. Rambouillet was an upmarket place with a good rail station and probably a commuter town for Paris. We (I) looked longingly into each of the *patisseries* at the marvellous *tartlettes* and *gateaux* on display from 3 euros each but I was promised hot chocolate croissants *chez* Katie (around 1.50 euros a bagful at Lidl) and they proved to be absolutely great warmed briefly in the oven. The sun tried to break through and persisted into the evening. Would it to be our last night in France?

The following morning after another cold night, but with the sun filtering through the trees, we took the opportunity to dump Katie's tanks and were on the road by 9.30 a.m. Paris our initial goal and the *'Périphérique'* our challenge of the day. Katie' satnav guided us onto the A12 and traffic remained busy then onto the A13 and before long we were into Paris and selecting '*Périphérique* North' alongside the Bois de Boulogne. The traffic continued to be busy but moving with '*fluide*' displayed on the overhead information signs. With many junctions for joining traffic I felt the inside lane was the worst place to be, then after sustained concentration and an hour from our start, we left the '*Périphérique*' heading for Charles de Gaulle airport. The weather wasn't being kind with intermittent dark clouds, rain spots then sunny breaks, but eventually showers and spray ensued. We made continual good progress. The A1 was busy with trucks but as we joined the A26 they melted away and all was quiet. We commented on this when we arrived six months previously. We made Calais by 2.30 p.m. and pulled into the '*sans billet*' car park and bought a ticket for 306.50 euros – cheaper than when we had booked ahead through the Caravan Club and were scheduled for the 3.30 p.m. ferry. Everything appeared to be going smoothly but then the lady from the office came running after us – Gail had left her passport behind.

We pulled into the embarkation lane, the sour-faced girl in the booth didn't say a word other than 'passports' – she returned them with instructions to join lane 20 and a ticket for the 3.30 p.m. ferry. We pulled forward to the customs and police who were curious to see inside but didn't search intensively for illegal aliens as I could have hidden four under the bed and twenty in the lockers.

We pulled into our designated lane. The ferry in front of us departed and we were left alone. All the trucks were waiting in the next section so I walked over and the official didn't know why we were in lane 20 and directed us to park with the coaches. Eventually we were signalled on, the marshal saying '*autobus, voiture*?' Then shrugged his shoulders – what were we?

Sitting on the ferry provided a short time for reflection, I was going to miss what we had been doing and a door was closing.

Thursday March 25 and we were back on UK soil. We had the usual problem disembarking – did we go trucks/freight or cars/coach – we decided coaches ('there may be a height barrier'). This time the English officials stopped us and again wanted to look inside. The officer alighted and told his colleagues how fantastic Katie was then they spotted Boris behind, not realising he was attached, with mutterings 'What's that idiot in the car doing in the coach lane?' Only when Boris disappeared with us did they realise the situation.

We were out of the docks and Katie's UK satnav took us straight to the Folkestone campsite. It felt a bit weird driving on the left.

Because of a failure to reset our clock, Friday saw us rising surprisingly early. We drove back to Wickford experiencing the usual traffic jams on the M25 to make us feel at home then one roundabout and 300 yards away from our destination Boris's alarm started and continued as we drove into the compound where Katie was parked. I pulled the battery terminal but he still managed to keep squealing. It was just what we needed at the end of our 3,800-mile journey. I started to get all the cables out to charge his batteries again to get the alarm to reset when some jobsworth came over to tell me I had parked Katie in the wrong place. Welcome to Britain.

The next day I sat down with my computer that had been progressively failing from Bizanet and a sinking premonition that its death was imminent but like a

Hollywood hero's last lingering scene it flickered back into life long enough to spasmodically transfer all my photos and writing to my newer laptop before its screen went into a spin and finally faded like the hero's last flicker of his eyes.

'You know, other people might be interested in what we did,' I reflected.

'I might write a book.'

Postscript

The legalities for owning and driving an RV were based on a personal interpretation of the information available at the time. As legislation often changes do not place any reliance on this. Do your own research and consult with those knowledgeable in the field and not those trying to sell you an RV.

Similarly, do not rely on the campground information without checking. Many of the Spanish coastal sites have made way for concrete warrens for holidaymakers and retirees.

Enjoy your travel.

Lightning Source UK Ltd.
Milton Keynes UK
UKOW03f2313160514

231847UK00001B/9/P

9 781782 999591